MARS

MAKING CONTACT

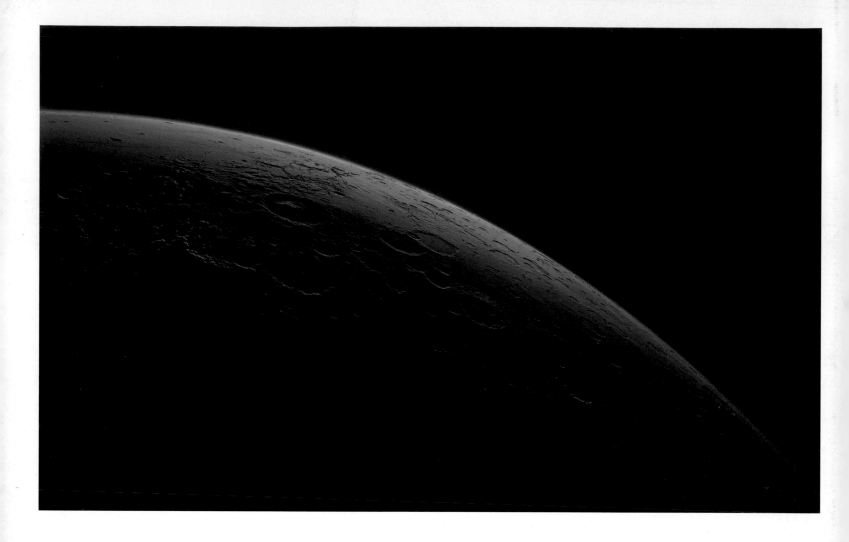

This book is dedicated to Susan Holden Martin, my favorite Martian, a trusted friend and valued ally through thick and thin.

THIS IS AN ANDRE DEUTSCH BOOK

Published in 2016 by André Deutsch
An imprint of the Carlton Publishing Group
20 Mortimer Street, London W1T 3JW

Text © Rod Pyle 2016
Design © Carlton Books Limited 2016

A CIP catalogue for this book is available from the British Library.

ISBN 978-0-233-00492-1

Printed in China

10 9 8 7 6 5 4 3 2 1

ABOVE: Dawn comes to Gale Crater, at center just below the cracked terrain. The magnificent Mount Sharp, created by sedimentation, can be seen in the crater's center. This image was created by digitally processing images from Mars Global Surveyor.

MARS

MAKING CONTACT

ROD PYLE

ANDRE
DEUTSCH

CONTENTS

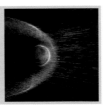

FOREWORD

From the very beginning when humans looked up at night, we knew Mars was a special place. It has played a unique role in our history, our scientific thinking, and in our dreaming of living on another planet. At the start of the space program when the US and the Soviet Union were just planning on sending humans to the Moon we were sending spacecraft to Mars.

But make no mistake. Mars is hard! A large number of missions tried and failed over the years. We always learned valuable lessons from each one of our robotic missions, by taking incremental steps first with flybys, then orbiters, then landers, ultimately leading to our successful rovers. Incremental steps like these will eventually allow humans to go beyond our planet's gravity and onto Mars. Rod captures this so nicely, building up the story with the set of robotic missions he has selected to discuss.

Our robotic explorers are gathering data to help us understand both the present and the past Mars. In August 2012 the Curiosity rover landed in an ancient river bed, determined the age of the surrounding Martian rocks, found evidence the planet could have sustained microbial life, took the first readings of radiation on the surface, and showed how natural erosion could be used to possibly reveal the building blocks of life protected just under the surface. Our recent Mars data has revealed that Mars was more like Earth in its distant past with rivers, lakes, streams, a thick atmosphere, clouds and rain and, perhaps, an extensive ocean. Although today Mars is rather arid it is now believed by our scientists that there are large amounts of water trapped under the surface of Mars and under the carbon dioxide snow of its polar caps. Water is the key that will enable human activity and long-term presence on Mars.

The entire world lived Curiosity's "seven minutes of terror," the time it took to go from the top of the Mars atmosphere to sitting safely on the surface. But those fleeting moments are nothing but the culmination of many struggles and triumphs along they way of all the missions that came before it. Rod's book captures the challenges we face from the imagined idea, to the planning, design and implementation, and finally the science of a Mars mission.

I have read many of Rod's books and have been impressed by his understanding of the missions as well as his ease in telling the story. His book offers a keen insight and intellectual analysis of our full exploration of Mars. His behind-the-scenes look engages the readers and gives them a glimpse of the journey in space exploration. These are truly exciting times in human history, our unfaltering vision, and our fearless journey to Mars.

Rod provides us all an overview of our robotic exploration in beautiful style taking us all on this grand adventure.

Jim Green, NASA's Planetary Science Division Director

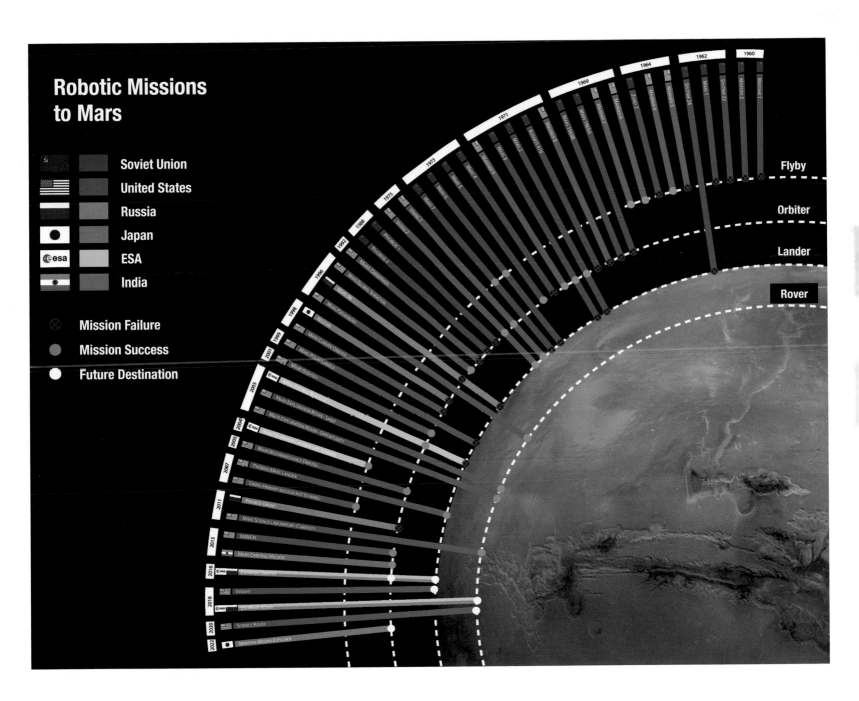

Robotic Missions to Mars

Soviet Union
United States
Russia
Japan
ESA
India

⊗ Mission Failure
● Mission Success
○ Future Destination

Flyby
Orbiter
Lander
Rover

1960 — Marsnik 1
Marsnik 2
Sputnik 22
1962 — Mars 1
Zond 2
Mariner 3
Mariner 4
1964 — Zond 2
Mariner 6
1969 — Mariner 7
Mars 1969A
Mars 1969B
Cosmos 419
1971 — Mariner 8
Mars 2
Mars 3
1973 — Mars 4
Mars 6
Mars 5
Mars 7
1975 — Viking 1
1988 — Phobos 2
Phobos 1
1992 — Mars Observer
1996 — Mars Global Surveyor
Mars 96
Mars Pathfinder
1998 — Nozomi
1999 — Mars Climate Orbiter
Mars Polar Lander
2001 — Mars Odyssey
2003 — Mars Express / Beagle 2
Mars Exploration Rover - Spirit
Mars Exploration Rover - Opportunity
2004 — Rosetta (Primary Mission Not to Mars)
2005 — Mars Reconnaissance Orbiter
2007 — Phoenix Mars Lander
Dawn (Primary Mission Not to Mars)
2011 — Phobos-Grunt
Mars Science Laboratory-Curiosity
2013 — MAVEN
Mars Orbiting Mission
2016 — ExoMars Orbiter
2018 — InSight
ExoMars Rover
2020 — Science Rover
2022 — Martian Moons Explorer

A FLICKERING ORB OF RED

Mars has always held a special place in the human psyche. Unique among the planets, with its dull red appearance and unusual motions in the night sky (it is the only celestial object whose course through the stars regularly alters), it has inspired awe and fear across the centuries. For thousands of years, most cultures have associated the planet with fire, blood, and violence. Mars evolved from being an otherworldly entity—a god of war—to a place.

THE TELESCOPE CONFIRMED THAT THE PLANETS were different from the stars, not just via their wandering motions, but also because they were clearly something other than twinkling points of light. They appeared as disks and had physical dimensions. These were actual spheres, suspended within the heavens.

Mars was long thought to be a near twin of planet Earth. Along with Venus (which was believed to be a warm, swampy place), Mars was known to be a neighboring world, one not too much farther from the sun and similar in its origins. Perhaps it was just another Earth, with its red soil hosting hardy cold-tolerant plants, strange new animals and, many thought, advanced beings that were not too dissimilar from us? The truth would actually turn out to be far less appealing.

Since the discoveries of the early space age, the scientific perception of Mars has changed from it being a potentially hospitable twin planet to a hostile, yet still familiar, frozen desert. Mars is similar to Earth in that the same rules of geology and physics fundamentally hold true, but the planet is in fact a dry, cold, windswept wilderness of rock and sand. Its atmosphere is so thin that water evaporates in a flash whenever it appears on the planet's surface. Ferocious winds occasionally scour the landscape with their cargo of erosive dust and sand. A relentless cycle of temperature changes—from cold to deep-freeze—break down the endless vista of rock that protrudes out of the red desert. This is the real Mars: a cold and forbidding place that we can now explore in ever greater detail through the use of increasingly advanced technologies.

Yet the planet still has a grip on the human psyche that is more than the sum of its dry rock and sand. Alone in the solar system, Mars represents a place that we can understand, and think of visiting and perhaps one day colonizing. It is the sole survivor of the early violence of the solar system, beckoning us as the only place we can soon reach off-Earth that offers some refuge from the empty vacuum of space that surrounds us.

Mars is a world, a destination, and potentially, another home for humanity.

THE ANCIENT SKY

If, one warm summer evening, you should find yourself far from the cloying light that pervades the night skies of the twenty-first century, take pause to look skyward. If you have not done this in recent years, wait a moment for the grandeur of the night sky to soak in. As you look into the heavens, you will see increasing numbers of stars… about three thousand are visible on a clear night. Soon a few of those points of light will look different— brighter, more pronounced, warmer in color, and twinkling less than the thousands of points of light that surround them. These are the planets of our solar system. Venus, Jupiter, Saturn, and Mars are easily seen with the naked eye. These objects move a little differently than the stars—hence the name "planets," derived from the Latin *planates*, or wandering stars. The ancients went through great pains to determine why these bright objects moved in unique motions relative to the others. Most of their conclusions were flawed, and a host of fantastic scenarios were developed to explain the phenomena. Among those puzzling objects, none were so vexing as Mars.

As you savor this unique look at the night sky, try to let your mind drift free of the shackles of knowledge acquired since the

OPPOSITE: Mars as seen from the Viking 1 orbiter in 1976. This is a compilation of multiple images, and has been processed to enhance dark surface features. Viking returned the first color pictures of Mars, which showed stunning levels of detail.

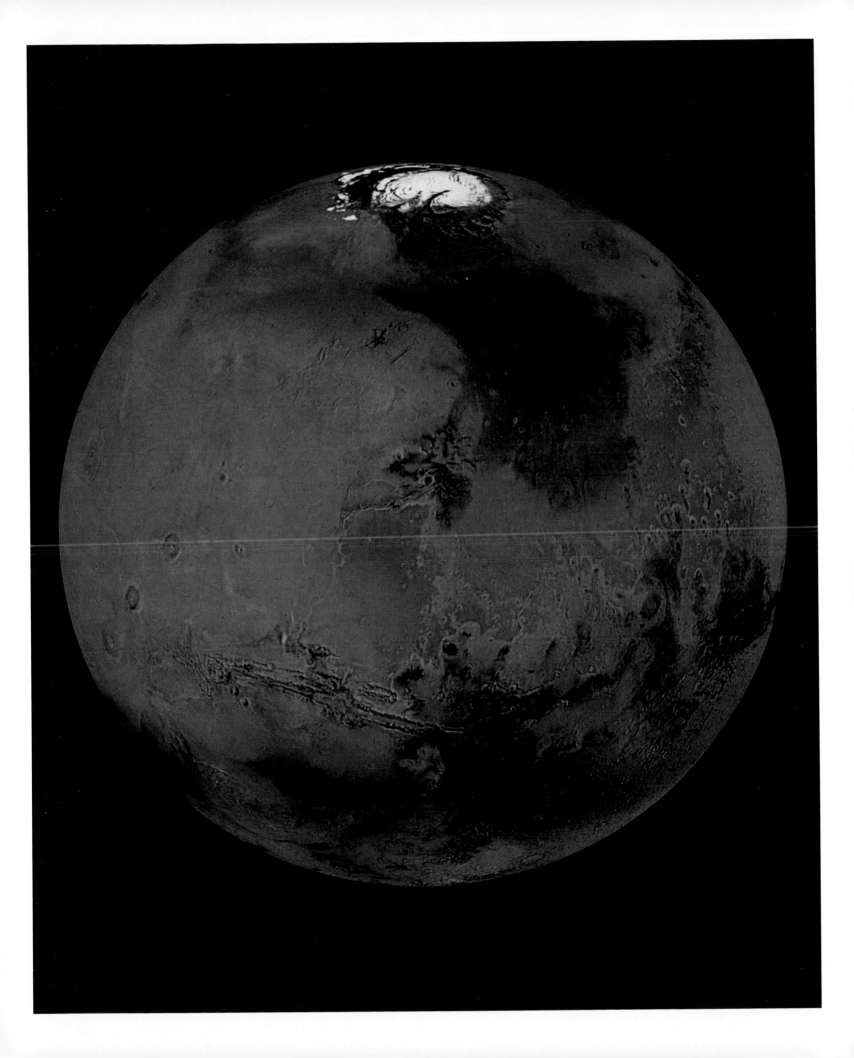

Age of Reason. Try to see the night sky as an ancient Egyptian, or a Greek, or a court astronomer of the Han dynasty in China would have done. In their time, the only light you would find at night would be that from the flickering candles of your home or the torches lighting the center of the village. However, outside that small circle of civilization, darkness prevailed, and the stars above would be your sole companions. Viewed in this way, against that twinkling blanket of stars, the planets would have appeared brighter and somehow more welcoming.

Except perhaps for one. At the best of times, Mars can be a disturbing sight, depending on your interpretation. Most profound is its color: depending on the atmospheric conditions, the planet can appear to be an almost blood-like red and, to a lesser extent, fiery in nature. Blood played such a visceral part in ancient societies (admittedly, people saw a lot more of it in day-to-day life than we do today), and life was sufficiently fleeting, that the vision of a drop of bright red in the sky would have been unnerving to many people looking up into the night sky in wonder.

Upon longer observation, over the course of a few months, Mars did something odd once every two years. Ceasing its normal progression across the sky, which changes the planet's position relative to the stars slightly night after night, Mars would suddenly reverse direction for a time, and then resume its normal path. This was puzzling. Today we know that this phenomenon is due merely to the arrangement of the orbit of the Earth inside the orbit of Mars. Every two years, the Earth catches up with, and then overtakes, Mars' larger, slower orbit. However, to the ancient observer, this must have been a curious, if not worrisome, change in the planet's behavior. For this reason, Mars was an outsider among the other planets.

Various societies came up with different stories to explain why Mars looked and behaved the way it did, most of which involved some form of carnage. Here is a selection from around the ancient world:

India—Mangal

Ancient East Indian myths surrounding Mars treated the planet as a god and named it Mangal, Mangala, Angaraka, or Bhauma. This god was considered to be "auspicious, like burning coal, and the fair one." This deity was organized and efficient, but also argumentative and combative, if not downright warlike.

Mesopotamia—Nergal

Ancient Mesopotamia equated Mars with the god Nergal, representative of war and pestilence. Nergal is presented as a ferocious lion or a large bird, or sometimes a combination of both. Over time, Nergal evolved into the overseer of the dead, who existed in the underworld.

Egypt—Horus the Red

In ancient Egypt, Mars was associated with the god Horus, generally considered the rebirth of Osiris (there are a number of versions of this myth), and also known as Horus the Red or Har Deshur. Early in the Egyptian pantheon, Horus went through a number of incarnations, alternately representing the harvest, the sky, war, and hunting. His head was that of a bird, as often seen in hieroglyphs. In another incarnation, Mars was the "good warrior," protecting the common man as the god Anhur, who began life as a war god.

Greece—Ares

The ancient Greeks' god for Mars was named Ares. He represented the physical brutality of war, and was often viewed with contempt and even revulsion. Zeus, the king of the Greek gods, considered Ares to be the god who was the most hateful and repellent of his charges. Ares was regarded as hotheaded, impulsive, and dishonest, and generally occupied a necessary but disdained place in Greek culture. One account referred to Ares as "overwhelming, insatiable in battle, destructive, and man-slaughtering." In contrast, Athena, his sister, represented a steady military hand, firm leadership, and superior strategic ability.

Rome—Mars

The ancient Romans appropriated the general mythic framework of Greece, renaming the gods that they adopted in the process. Ares became Mars, and was elevated from a figure embodying necessary evil to someone to be admired and emulated. The very traits that had repelled the Greeks seemed to appeal to Rome, which found Mars' martial ferocity more aligned to their own bloody process of continuing conquest.

Western Europe—The Middle Ages

By the late sixteenth century, observers of the red planet had amassed a long list of specific traits attributed to Mars. "Mars rules catastrophe and war, it is master of the daylight hours of darkness of Tuesday, and the hours of darkness on Friday, its element is fire, its metal is iron, its gems jasper and hematite, and it rules the color red. Its qualities are warm and dry, it rules the color red, the liver, the blood vessels, the kidneys, and gallbladder as well as the left ear. Being of choleric temper, it especially rules males between the ages of forty-two and fifty-seven."

In later centuries, by the time Galileo and others began turning telescopes to the sky, Mars gradually became more of a "what" than a "whom." The Age of Reason took away a bit of the fun,

and drama, of man's relationship with Mars. That is, until the late nineteenth century, when astronomers such as Giovanni Schiaparelli, Camille Flammarion, and Percival Lowell began to intuit the idea of advanced life forms and massive canal-based engineering projects from swimming telescopic images. This construction of Mars as a place of intelligent empire would be almost as thrilling, and wrong-headed, as those misguided concepts of the ancients.

ABOVE: An engraving by Andreas Cellarius, c. 1661, depicting astronomers using early telescopes to explore the cosmos. Tools and parts at the bottom seem to indicate an additional device being assembled.

RED PLANET EMPIRE

By the time of the Renaissance in the sixteenth century, the planets had become places as opposed to deities, and this opened up entirely new avenues of thinking. In 1543, the Polish astronomer Nicholas Copernicus published an influential work entitled *De revolutionibus orbium coelestium* that finally placed the sun in its rightful place at the center of the solar system, and the planets in orbits around it.

AT THE END OF THE SIXTEENTH CENTURY, Johannes Kepler, a German astronomer, determined the nature of planetary orbits. A considerable amount of his efforts was focused on Mars and its bizarre retrograde motions in the night sky.

The Italian astronomer Galileo Galilei built some of the earliest telescopes used for astronomical purposes and in 1609 noted that the planets were disks as seen against the point-like fire of the stars. His instruments were small and ranged from about three-times magnification to a maximum of thirty. This was not enough to see any details on Mars, but did allow him to observe the phases of Venus, four of Jupiter's moons, and the rings of Saturn.

In 1659, during one of Mars' close approaches to Earth, Dutch

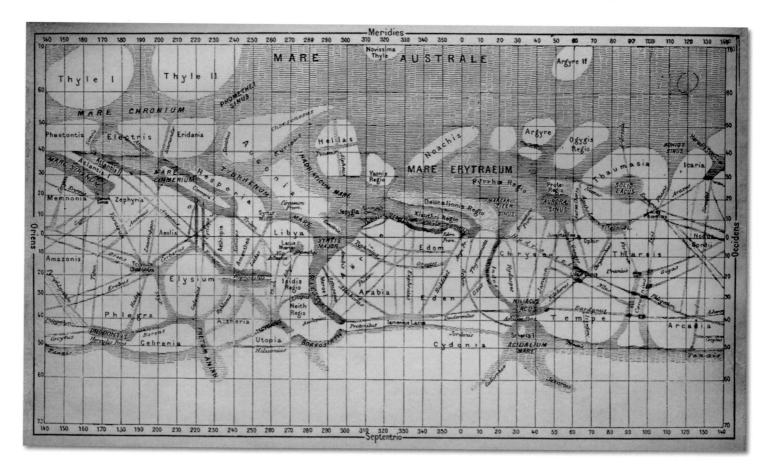

astronomer Christiaan Huygens used a telescope of his own design to make early drawings of Mars. These close approaches occur every two years and cut the distance between Mars and Earth from a maximum of about 255 million miles (410 million km) to as little as 33 million miles (53 million km). Huygens' simple Mars maps included a large dark spot, a region later named Syrtis Major. He noted that the dark area returned to view approximately every night, allowing him to perform rough measurements of the Martian day at somewhat over 24 hours. Huygens also noted a south polar cap a few years later. Mars was slowly giving up its secrets to the power of the telescope.

ADVANCES IN TELESCOPIC AND MAPPING TECHNOLOGY

Between 1777 and 1783, English astronomer William Herschel used his new and large aperture telescopes to make detailed examinations of Mars. Over his life, Herschel made hundreds of telescopes, including one nearly 50in (127cm) in diameter, by far a record size for the time. Using his state-of-the-art optics, Herschel was able to determine the axial tilt of Mars, or inclination of its axis of spin, which led him to surmise that Mars might have seasons not unlike those of Earth. His telescopes were sufficiently powerful that he was even able to note the slight scattering of light from stars passing behind Mars, and he concluded that there was an atmosphere of some kind surrounding the planet. Add this to an Earthlike day/night cycle, which was known to be twenty-four hours and forty minutes in duration, and Mars was beginning to look as though it had the potential to be a sister planet to Earth.

This idea became more ingrained in the nineteenth century. Improved telescopes made for skillfully drawn, greatly improved maps. At last clear features could be carefully charted on paper—at least that is what their creators thought. At the time, the best images were accomplished by pencil sketch, drawing what was seen in the eyepiece. Simple diagrams would be made night after night, and then combined later on into more detailed renderings. Astronomers such as Richard Proctor, an Englishman, created maps that clearly showed major surface features, as did the American Asaph Hall. The idea that the lighter, red regions on Mars were land masses and that dark areas were oceans was coming into vogue, but it was puzzling that these changed from one viewing season to another (the two-year close approach cycles). Others theorized that the darker areas they saw might be huge masses of plant life, such as forests, or ground vegetation.

In 1877, an Italian astronomer named Giovanni Schiaparelli conducted an intensive observing program during the favorable opposition of that year. Earlier maps had tended to have features named in the language and traditions of the observer's home country, and Schiaparelli chose nomenclature in keeping with Italian tradition, labeling major features in Latin. These names are still in general use today.

Schiaparelli's maps of Mars were works of art, rendered in fine detail, and with the hand of a skilled craftsman. Unfortunately, this kind of interpretative drawing, which blended weeks or months of observations, also introduced some subjective imagery. These maps included linear markings that connected some of the dark regions, which Schiaparelli called *canali* (Italian for "channels"), and this was soon translated into English as "canals." Whether this slight error in translation was an innocent oversight, a product of whimsy or an attempt at sensationalism is not known for certain, but in any case the effect was profound. Channels could be the result of some natural process, such as groundwater runoff—but canals would have to be the work of intelligent beings, at least according to the popular press of the time.

There was some resistance to the idea of intelligent beings inhabiting Mars but, in truth, nobody could really say. The best maps were approximations of visual observations as interpreted by the observer. Early spectroscopic observations, which split the light from a planet or star into prismatic colors and allowed for the determination of chemical elements, were imprecise. They did, however, seem to indicate the existence of water vapor in the Martian atmosphere, and while that atmosphere was understood to be much thinner than that of Earth, it was thought that it might be of sufficient density to support some form of life.

THE NEW MYTHOLOGY OF MARS

The more fanciful extremes of life on Mars were colorfully expressed by the French astronomer Camille Flammarion, who was also a spiritualist and early science fiction author. The boundaries of his skill set seem to have been somewhat blurry, as perhaps was his eyesight. In the early 1870s, he wrote in great detail about his own observations of Mars and his response to those from other astronomers. After a brief recap of Earth's continents, he said:

> "It is different with the surface of Mars, where there is more land than sea, and where the continents, instead of being islands emerging from the liquid element, seem rather to make the oceans mere inland seas—genuine Mediterraneans. In Mars there is neither an Atlantic nor

OPPOSITE: Giovanni Schiaparelli's map of Mars produced from sketches 1877–1886. Surface features were enhanced by Schiaparelli and depicted details, such as the *canali*, that were not really there. His maps formalized the use of Latin names for Martian features.

a Pacific, and the journey round it might be made dry-shod. Its seas are Mediterraneans, with gulfs of various shapes, extending hither and thither in great numbers into the terra firma, after the manner of our Red Sea…"

Flammarion goes on at great length, drawing increasingly unsupported conclusions. They are fascinating, if fanciful. But then, following a trend that emerges from his discourse, he begins to admit to his own editorializing, while arguing that it is justified.

"We speak of plants on Mars, of the snows at its poles, of its seas, atmosphere, and clouds, as though we had seen them. Are we justified in tracing all these analogies? In fact, we see only blotches of red, green, and white, upon the little disk of the planet; but, is the red, terra firma; the green, water; or the white, snow? Yes, we are now justified in saying that they are. For two centuries astronomers were in error with regard to spots on the moon, which were taken for seas, whereas they are motionless deserts, desolate regions where no breeze ever stirs. But it is otherwise as regards the spots on Mars."

Flammarion's wild ideas had a ready audience in the West. Spiritualist movements in Europe and the US were exploring and popularizing alternatives to traditional religion, and the possibility of intelligent life beyond Earth became part of some of their teachings. The idea that nearby planets such as Mars and Venus could be our distant cousins was just too appealing to ignore.

All the while, other astronomers stared intently into their eyepieces. Even with the fine telescopes emerging in the late nineteenth century, featuring large lenses and mirrors, and tremendous light-gathering ability, Earth's atmosphere was the constant enemy. Turbulent and fickle, the blanket of air that surrounds our planet causes any attempts to peer into space to be problematical. What is known in the astronomical trade as "seeing"—the quality of an image coming into a telescope due to the state of the air mass above it—can range from excellent to deplorable in a matter of minutes. On even the best nights, in a well-equipped observatory situated on a high mountaintop, Mars is so far away that it is a shimmering blob of red with a few dark shadings across its surface. To interpret these as rivers, seas, and continents certainly took a lot of imagination.

PERCIVAL LOWELL'S DREAMS OF THE RED PLANET

The mystical aura surrounding Mars permeated the psyche of a young American named Percival Lowell. He was the consummate example of the right man in the right place at the right time to take-

up the rallying cry for intelligent life on Mars. Harvard-educated, Lowell came from a wealthy Boston family and spent years in Asia in his youth, later writing popular books about Japanese culture. After returning to the US, Lowell nourished his fascination for astronomy by consuming many books on the subject, including at least one by Flammarion. Flammarion's ideas, combined with Schiaparelli's suggestive maps and observations, became a fixation for Lowell, and by 1894 he had decided to spend a substantial part of his personal wealth on coming to know Mars in more intimate terms. Lowell bought a mountaintop in the Arizona territory and built an observatory outside the town of Flagstaff, dedicated exclusively to observing Mars.

Lowell's extensive observations resulted in maps even more intricate than Schiaparelli's. He laboriously numbered each and every canal he saw through the eyepiece and created a vast system of interconnected features on the Martian surface, the sum of which gave rise to the ultimate expression of an argument for intelligent life on Mars. He wrote a series of very popular books, expressing increasingly intricate (and unlikely) ideas about what kind of society must inhabit Mars. These included detailed thoughts about the vast systems of civil engineering and planet-wide government that must be required to accomplish the great works of engineering he thought he saw through the eyepiece of his telescope.

In Lowell's mind, the Martians were in the last stages of trying to save their beleaguered planet. In a scenario not far removed from the fiction of H.G. Wells' Martians in his seminal tale of invasion, *The War of the Worlds* (which came out, perhaps not coincidentally, at about the same time as Lowell was publishing his own works), the canals of Mars were huge engineering masterpieces designed to transport water from the icy poles to the parched and arid equatorial regions. Lowell worked out the underpinnings of an entire civilization based on hazy smudges in the ever-wavering image of Mars in his telescope. It should be mentioned that he did so with a fair amount of scientific rigor for his time, and his books still make fascinating reading, even though the ideas within lead to conclusions that are generally opposed to reality.

In this carefully constructed Martian empire, Lowell noted that:

"the aspect of the lines [canals] is enough to put to rest all the theories of purely natural causation that have so far been advanced to account for them... first, the straightness of the lines; second, their individually uniform width, and third, their systematic radiation from special points..."

He argued against natural causes for these observed features and for an intelligent hand in their creation. In the early 1900s, he wrote:

"That Mars is inhabited by beings of some sort we may consider as certain as it is uncertain what these beings may be... Girdling the globe and stretching from pole to pole, the Martian canal system not only embraces their whole world, but is an organized entity. Each canal joins another, which in turn connects with a third, and so on over the entire surface of the planet. This continuity of construction posits a community of interest... and the thing that is forced on us in conclusion is the necessary intelligent and non-bellicose nature of the community, which [sic] could thus act as a unit throughout its globe."

Sadly, Lowell's observations, along with those of Schiaparelli and others, were distorted by the shortcomings of the available optics of the era, along with a healthy dose of fanciful thinking. Telescopes improved still further and other methods of investigation, including improvements in spectroscopy and radio telescopes, brought the idea of anything more advanced than rock-clutching lichen living

OPPOSITE: Percival Lowell at his 24-inch refracting telescope, purpose-built for observing Mars, in his observatory in Flagstaff, Arizona, 1914.

ABOVE: One of Lowell's maps of Mars, created by combining sketches made from numerous observations. Lowell amplified the linear features suggested by his predecessors.

RIGHT: A computer artwork of the cargo ship and glider designed by the German-US rocket engineer Wernher von Braun, as part of his proposal for a manned mission to Mars.

on Mars into doubt. There were likely no canals, no planet-wide governments and no Martians, though many of his readers and a few scientists remained hopeful despite the increasing evidence against life on Mars. However, in order to be certain whether there was life or not, we would have to go there—and thinking about spaceflight in general, and a manned journey to Mars in particular, began just a few decades later.

EARLY CONCEPTS OF SPACEFLIGHT

The Russian scientist Konstantin Tsiolkovsky wrote of manned spaceflight early in the 1900s, inspiring a generation of scientists and engineers. Of these, a young German engineer named Wernher von Braun was the first to tackle the problem with a full treatise, in the form of a book he completed between 1948 and 1952. Published first in German as *Das Marsprojekt*, and later in English as *The Mars Project*, the book outlined the projected requirements for this enormous undertaking. Included were intricate spacecraft designs, the number of rocket launches (950) to get the required mass to orbit, and the flight-time to Mars for the armada of ten spacecraft he envisioned. He designed winged gliders that would land crews on the relatively smooth Martian poles, who would then make an overland trek to the equatorial regions. The crew would depart just over a year after rejoining the orbiting fleet for a return to Earth. The projected crew numbered seventy men.

Von Braun's mission concept was published in *Collier's* magazine and outlined on some of Walt Disney's popular family television programs. In the late 1950s, with memories of World War II fading and the age of the atom upon us, anything seemed possible.

Then, in 1965, a tiny machine sped past Mars, smashing both the Victorian fantasy of Martian empire and von Braun's mission designs to red dust.

THE SPRINTER: MARINER 4, FIRST TO MARS

The space age began not with a bang, but with a beep. In October of 1957, the Soviet Union launched Sputnik (Russian for "moon") into orbit around the Earth. This simple satellite, the world's first, was sent into space atop one of the USSR's new nuclear missiles, an R-7 ICBM. Sputnik glided into orbit, doing little more than beeping every few seconds via a small radio transmitter. But that was enough—the radio signal was heard in receivers across the world, and the West was spurred to increased action.

AS THE UNITED STATES STRUGGLED TO MATCH THE SOVIET accomplishment, scientists at places like Caltech were thinking beyond orbit, about ways to get a better look at the planets. Once the US had matched the USSR's achievement, the newly formed NASA was able to split its energies between getting a man into space (an endeavor in which the Soviets also succeeded first), and designing robotic probes to look at Earth's two closest neighbors, Venus and Mars. Of course, the Soviets had the same planetary targets in mind.

THE EARLY MARINER VOYAGERS

NASA's first attempts to reconnoiter Mars were called Mariner. The Mars-bound version of the spacecraft had a flat-sided, ring-shaped chassis to which cameras, radio dishes, and solar power panels were affixed, configured for the mission at hand (an earlier design was sent to Venus with a different layout). The mariners were almost skeletal in appearance, with the hardware created to exist in the vacuum of space and hardened to the radiation and harsh temperature swings of that environment. It was a light but robust

design. The Soviets followed a somewhat different engineering path, using large and heavy pressurized hulls, which had the virtue of adding much-needed protection to the delicate components at the cost of weight. But with the larger Soviet rockets continually outpacing their US cousins in propulsive power, they could afford to fly the extra mass. This also allowed them to design more elaborate missions, including impact probes and primitive landers, though the majority of them failed to work in practice.

The first two Mariners were sent to Venus. Mariner 1 was launched in July 1961, but failed shortly after lift-off. Mariner 2 enjoyed success after an August 1962 launch and became the first spacecraft to fly past another planet just under four months later, sending back simple but important measurements. However, there were no pictures, as it did not carry a camera. The Soviets began sending their Venera spacecraft to Venus in 1961 as well, but were foiled in their attempts when the first three missions lost contact with Earth before reaching the planet. (They would not succeed until Venera 4 managed to reach Venus in 1967.)

With the success of Mariner 2, Mars was in the crosshairs of the competing superpowers. But the voyage to Mars made that to Venus look easy. The journey took about twice as long, and sent the spacecraft much farther from its Earth-bound controllers. Heading to Mars also took the spacecraft away from the sun, so the power-hungry electronics of the day needed larger solar panels in order to generate sufficient power.

The Jet Propulsion Laboratory (JPL) in Pasadena, California, was NASA's west coast base of operations for robotic exploration. At this point, both the US and USSR were splitting their efforts between human exploration and robotic probes. President John F. Kennedy's public challenge—to beat the Soviet Union to a manned moon landing—had been issued in late 1961 and drained the majority of NASA's funding. Much of what was left was sent to the planetary exploration program, now aimed squarely at Mars.

Unlike any other NASA center, including the famed Cape Canaveral in Florida (later named the Kennedy Space Center) and the Manned Spaceflight Center in Houston (later named the Johnson Spaceflight Center), JPL was not run directly by NASA. For a variety of practical and historical reasons, NASA hired the California Institute of Technology, also in Pasadena, to manage JPL for them. It was a fortuitous arrangement—the professors at Caltech provided a high-order brain trust of planetary science and engineering know-how.

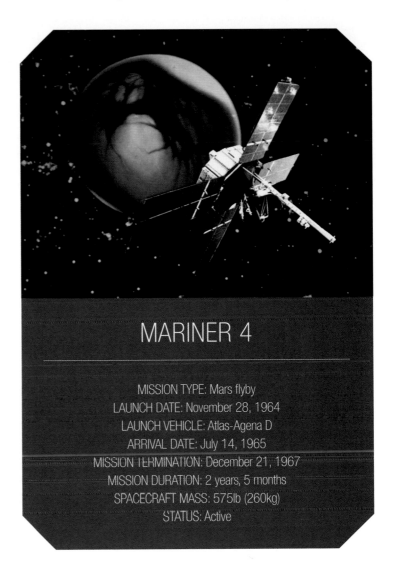

MARINER 4

MISSION TYPE: Mars flyby
LAUNCH DATE: November 28, 1964
LAUNCH VEHICLE: Atlas-Agena D
ARRIVAL DATE: July 14, 1965
MISSION TERMINATION: December 21, 1967
MISSION DURATION: 2 years, 5 months
SPACECRAFT MASS: 575lb (260kg)
STATUS: Active

DEVELOPING THE TECHNOLOGY

A team of these professors was assembled to ponder a redesign of the basic Mariner spacecraft intended for a Mars flyby. In the early 1960s, orbiting another planet was not on the agenda; simply reaching Mars would be a major challenge, so the planetary missions of that decade were based on simpler flyby trajectories. As Mariner neared Mars, it would burst into action, taking measurements and snapping photos as it swung by the planet at interplanetary speeds.

The team of Caltech professors, working with JPL engineers, got busy building Mariners 3 and 4. The spacecraft was a new design that would carry instrumentation but, like the the Mariner Venus probe, would not include a camera. Upon reviewing the proposal, three members of the Caltech team protested: Robert Leighton was a professor of physics with a long background in visual astronomy; Gerry Neugebauer was a professor of geology; and a young Bruce Murray, a recent addition to the faculty, who would later become the director of JPL. Collectively, they felt that the visually complex surface of Mars merited the inclusion of a camera to send images back to Earth. Venus, with its perpetual cloud cover, had not been deemed photographically interesting, so other instrumentation had sufficed for those missions. However, Mars as seen through the telescope was a constantly shifting panorama of light and dark, red and gray areas. What did these represent? Spectrographic and other data would provide answers about the makeup of whatever was on the surface, but matching these data with photographs would be critical. Leighton also suspected that the public, with so many of their tax-dollars being spent on space exploration, would be excited to actually see the planet close-up.

This was all very well, but in the early 1960s television cameras were the size of a dishwasher and extremely heavy. They also used brittle glass Videocon vacuum tubes for imaging, which created heat and were extremely power hungry. All these factors worked against taking such a device on a spacecraft. Consequently, Leighton's team created a low-weight, low-energy design from scratch. The final design was slow, fragile, and created black-and-white images of just two hundred lines of resolution, each only two hundred pixels wide. But the tubes were comparatively robust, and worked in tests that simulated flight conditions. Other instrumentation included a magnetometer (for measuring magnetic fields), a dust/micrometeoroid detector, a cosmic-ray telescope, a radiation detector, a solar plasma probe, and a handful of other devices. Power was provided by four folding solar panels, with a total of 28,244 solar cells arrayed across them.

Intriguingly, and only on this first Mars-bound Mariner design, at the end of each solar panel was a flat, fan-shaped "petal." These were steerable vanes, and designed to aid in maneuvering and stabilizing the craft using the scant pressure of the solar wind, charged particles that stream from the sun. While they had a small effect, the vanes were not deemed to be worth the weight and added complexity, and were dropped after the missions of Mariners 3 and 4.

Beginning with the Mariner program, NASA (and, in general terms, the Soviets) wisely decided to build two spacecraft for each mission. The thinking was that spaceflight was so new, risky, and expensive, that it would cost less in the long run to build two at a time rather than to replace one if it failed. In many cases during the first decade of interplanetary spaceflight, the backup spacecraft ended up becoming the only spacecraft when one failed. Right at the start, Mariner 1's rocket malfunctioned at launch, leaving Mariner 2, its virtual twin, to complete the mission.

The same was the case with the first Mars-bound Mariner probes. On November 5, 1964, Mariner 3 ascended from Cape Canaveral, Florida, on an Atlas booster. The rocket reached orbit (not always guaranteed with the then-troubled Atlas), and Mariner

ABOVE: Mariner 4 prior to launch on November 28, 1964. The rocket is an Atlas booster with an Agena upper stage.

3 prepared to release the nose-cone shroud. This was a two-piece metal fairing that would split open and drift away from the Mariner spacecraft before the probe headed off to Mars… but it did not. The fairing was restrained by a metal band that was supposed to release the two halves once the rocket got into orbit, but the band stuck fast. Engineers could only watch as the first Mars probe slowly ran down its batteries, unable to deploy its solar panels or maneuver. It died a slow, energy-suffocated death in space.

Robert Leighton

Professor of Physics,
Caltech

Robert Leighton had been a professor of physics at Caltech for over a decade when he was asked to participate on the Mariner 4 mission. "For reasons that I'm not quite clear about, but possibly having something to do with the fact that I had worked on sun and the planets... I was dragooned by Bruce Murray and Gerry Neugebauer into participating in the Mariner 4 television experiment." Up until this time, NASA had planned to fly past Mars without a camera. "There had been no reasonable proposals for a photographic component of the mission, for television, for pictorial work." And that simply seemed unacceptable.

After a lot of trial and error, Leighton's team was able to build the first TV camera to go into deep space. He recalls seeing the first images from Mars... and keeping the news quiet for a while: "We knew there would be craters and so forth. And yet, the fact that craters were there, and were a predominant land form, was somehow surprising!" He was not quite as thrilled with the image quality, however. "The pictures were of such—I can't say poor quality, but at least, the [technical] limitations were so severe... that we waited a week or more, after we knew there were craters, before any kind of official announcement was made."

But once the images were sent to the media, they were a sensation. Leighton recalled in particular one letter he got from an Oregon dairy farmer who had seen those first images. It said: "I'm not very close to your world, but I really appreciate it, keep it going."

"I thought that was kind of nice," Leighton later commented.

Once the designers understood what had gone wrong, they performed a rapid redesign of the fairing (among other things, switching from metal to fiberglass) in what was surely a record at the time of three weeks. It was literally installed on the launchpad. Just weeks after the failure of its sibling, Mariner 4 lofted out of the Cape on November 28, 1964. Upon reaching orbit, the shroud separated as designed and the spacecraft left orbit and successfully headed off to Mars.

MISSION: MARS

However, leaving Earth was just the beginning of the drama. To navigate to Mars, the craft had to find its position relative to the cosmos. To accomplish this, two photocells were installed to track objects of known brightness. About a half hour after leaving orbit, Mariner was commanded to orient one of these tracking sensors toward the sun, an easy target to find. Then it would rotate around that axis, searching for a known star—in this case, the bright star Canopus. However, the sensor was not all that discriminating, and it took a full day with a half dozen other stars fleetingly in the crosshairs, before Canopus was identified and locked-in.

For six weeks this navigational drama repeated itself every few days, with the spacecraft repeatedly losing track of Canopus. If it could not reacquire the guide star, navigation would become pure guesswork. Time and again the star tracker drifted off Canopus, and time and again controllers would initiate a search to reacquire.

Analysis of the problem seemed to indicate that the tracker was operating properly—so why was Mariner continuing to lose its lock on the star? Eventually, after a lot of late-night sessions and deeply intuitive problem solving, the engineers realized what was happening. When the spacecraft had separated from the upper stage of the Atlas rocket, a small cloud of paint flakes, dust, and dirt carried aloft within Mariner's protective shroud was also expelled. These flakes, dust, and dirt were now surrounding Mariner 4, since there was no air in space to disperse them. If a sufficiently large fleck of paint drifted past the star tracker, the photocell would see it as a bright star and attempt to follow it rather than continue to stare at the dimmer Canopus. In effect, Mariner 4 had its own constellation of mini-stars flying in formation. A new set of parameters concerning the sensitivity of the Canopus tracker were transmitted to the spacecraft and its behavior was improved for the remainder of the mission.

Other than this navigational issue, the seven-month cruise to Mars was fairly uneventful. There was one instrument failure along the path to Mars—a radiation-sensing tube ceased to function a few months after launch. However, such incremental challenges

LAB·ORATORY JAN 1965

JET PROPULSION LABORATORY · CALIFORNIA INSTITUTE OF TECHNOLOGY · PASADENA, CALIFORNIA

IMPATIENT WORLD WAITS TO "TUNE-IN" ON MARS

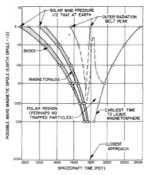

As Mariner IV races towards its July encounter with Mars, encounter planning continues a space at JPL Pasadena and at the far-flung elements of the JPL Deep Space Net. When Mariner IV passes within 5700 miles of the Martian surface on July 14, its scientific instruments will provide man with his first close-up look at the Red Planet.

Near-Mars Cruise Science Observations. The Mariner scientific instruments have been discussed in detail in previous Lab-Oratories, however, some comment may be made on their expected behavior as the spacecraft nears Mars. Figure 1 summarizes the expected performance of the cruise science instruments (cosmic dust detector, cosmic ray telescope, ion chamber, magnetometer, plasma probe, and trapped radiation detector) during the Mars flyby. If the region surrounding Mars is similar to that near Earth, the detection of magnetic effects and associated trapped particle densities is determined largely by the strength of Mars' magnetic dipole. Possible Mars magnetic dipole strengths compared with that of Earth are shown on the vertical axis. The horizontal scale shows the Pacific Daylight Time at which the event can be expected to occur at the spacecraft. The various effects expected near Mars are plotted as a family of lines, each of which represents the expected time of occurrence of that particular event, assuming various magnetic moments. The shock area represents the magnetohydrodynamic shockfront where an abrupt change in the plasma spectrum occurs as the solar plasma impinges on the Mars magnetic field. The magnetopause represents the transition region between the interplanetary magnetic field and the planetary field.

Near-Mars effects would be observed in order by the plasma probe, magnetometer, and trapped radiation detector. Some changes in the cosmic dust background may also be observed due to possible dust belts in orbit around Mars.

The investigators responsible for the cruise science experiments will be present at JPL Pasadena during the encounter operations and may be able to give a preliminary summary of their observations to the public on the day after encounter.

The Encounter Sequence. The encounter sequence consists of a series of spacecraft events required to properly position the television camera and record and playback television pictures of Mars. The events will be initiated either by command from Earth or by signals from equipment on board the spacecraft, depending on the options chosen by project management. The nominal encounter sequence includes:

(1) turn-on of encounter equipment about 9 hr before closest approach.
(2) acquiring the planet and stopping the planetary scan platform (Fig. 2) in the proper position for television pictures of Mars.
(3) recording the television pictures on the dual-track video storage tape recorder.
(4) stopping the recording sequence after the video tape is filled with pictures (about 20 minutes before closest approach).
(5) turning off the encounter equipment.
(6) playing back the television pictures stored on the video tape.

Closest Approach. At 1802 PDT on 14 July, Mariner IV will make its closest approach to Mars, passing within 5700 miles of the surface.

Occultation. About 1 hr after closest approach to Mars, the Mariner spacecraft will disappear behind the planet as seen from Earth and remain hidden (occulted) for about 50 minutes before emerging again. Sophisticated equipment at DSIF stations in California and Australia will record changes in the spacecraft radio signal caused by refraction (bending) by the Martian atmosphere and ionosphere. These data will be used to construct a more accurate model of the Martian atmosphere for use on future missions.

Picture Recovery. Picture playback will begin at about 0453 PDT on July 15 when the data encoder is transferred to Mode 4 (8-1/2 hr of picture data followed by 2 hr of engineering data).

The first picture will be received by the Johannesburg, South Africa DSIF station and transmitted via teletype to the Space Flight Operations Facility (SFOF), where the data will be processed into a television picture 200 elements square. If all goes well, the first close-up television picture ever taken of the planet Mars may be available for public viewing within 24 hr after closest approach.

NOTE: Video and audio coverage will be provided for JPL employees on July 14 in 180-101 and in the main cafeteria. The Von Karman Auditorium is reserved for the press. Details and schedules for encounter will be provided on a Mariner-Mars (green sheet) prior to encounter sequence.

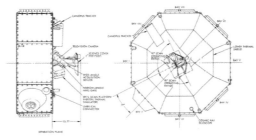

Fig. 1

Fig. 2a – The planetary scan platform is shown in a side view with the science cover deployed, revealing the wide and narrow angle planet sensors and the TV camera.

Fig. 2b – A view of the bottom of the Mariner spacecraft showing the 180 degree field through which the scan platform can move in search of the planet.

8   9

OPPOSITE: JPL's internal magazine, *Lab-Oratory*, was quick to dispense news of Mariner 4's mission to the magazine's audience. This front cover dates from January 1965.

ABOVE: "Impatient World Waits to 'Tune-In' on Mars". The headline of this article captures the public fascination with Mars at the time of the Mariner voyages; few details were overlooked while imparting technical detail to JPL's readership.

RIGHT: "Sequence of Events for Mariner IV". This short article is about spacecraft command and timing issues for the flight, including the "CC&S," the Central Computer & Sequencer, which sported a then-impressive 256 words of ability.

SEQUENCE OF EVENTS FOR MARINER IV

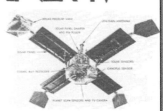

On January 3, 1965, the CC&S will transfer the spacecraft Data Encoder from the 33-1/3 bits per second to 8-1/3 bits per second. At this time the spacecraft will have travelled far enough from the Earth so that the signal received from its radio subsystem will no longer be strong enough to support the higher information rate. On March 4, 1965, the CC&S will command a transfer from the omni antenna to the high gain antenna in order to maintain sufficient signal strength for the balance of the mission.

As the spacecraft moves around the Sun toward its encounter with the planet Mars, the position of its star tracker will change relative to the star Canopus. As a result the angle of the Canopus tracker view window relative to the spacecraft - Sun line will have to be updated periodically through the mission. These cone angle updates occur on February 27, April 2, May 7, and June 14, 1965. The capability also exists to update the cone angle by ground command.

On July 14, 1965, as the spacecraft approaches Mars, the CC&S will command the turn-on of encounter science. At this time the scan cover will drop, and the scan platform will begin its search for the planet. When the platforms wide angle sensor acquires the planet, the scan subsystem will stop searching, and begin to track the planet. As soon as the spacecraft is close enough to Mars, sometime early on July 15, a narrow angle sensor will see the planet and command the start of the picture recording sequence, which requires some 25 minutes to record up to 22 pictures.

Within an hour after the closest approach to Mars, the spacecraft will fly behind the planet. The radio signals broadcast through the atmosphere of the planet are expected to yield information on the nature of the Martian atmosphere. Approximately an hour later, the CC&S will command the turn-off of encounter science, and 6-2/3 hours after that will command the start of the picture playback. To receive a single playback of all pictures at SFOF will require nearly ten days. After the playback is complete, the spacecraft will continue to return to earth scientific data as long as communications can be maintained.

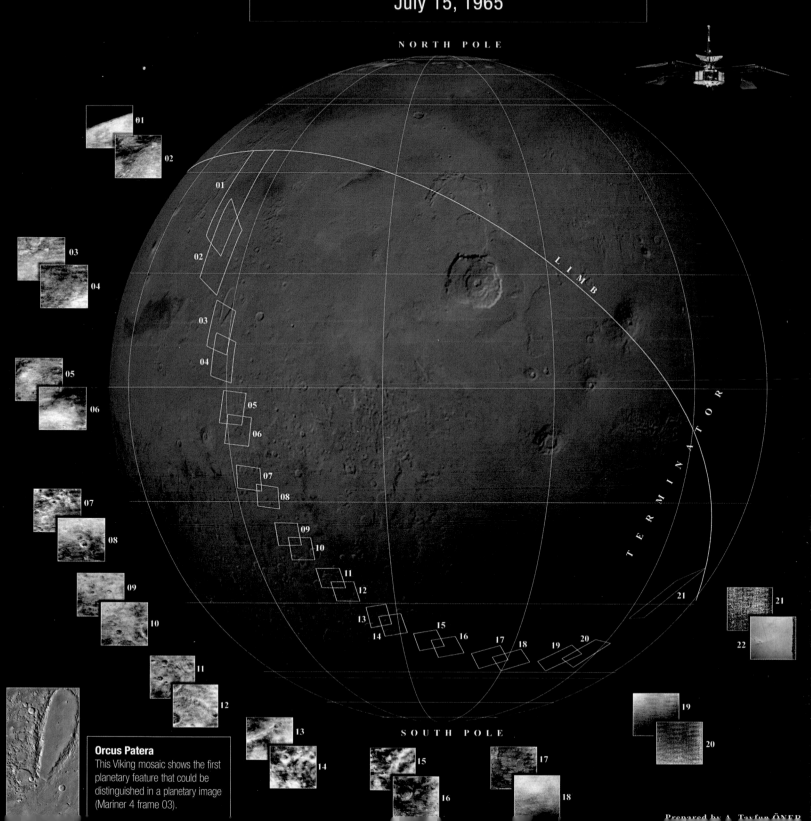

Mariner IV Mars Encounter Imaging Geometry
July 15, 1965

NORTH POLE

LIMB

TERMINATOR

SOUTH POLE

Orcus Patera
This Viking mosaic shows the first planetary feature that could be distinguished in a planetary image (Mariner 4 frame 03).

Prepared by A. Tayfun ÖNER

were to be expected—after all, this was only the second time an interplanetary probe had successfully left Earth and continued into deep space. It was also the first time that anything had succeeded in heading toward the outer solar system. Mariner 4 was operating on the edge of the unknown.

MISSION ACCOMPLISHED

On July 14 and 15, 1965, Mariner 4 flew past Mars. As the planet loomed out of the darkness, the probe's instruments were activated to take their readings on the one-pass opportunity—there would be no second try. Just prior to the spacecraft's closest approach to the planet, the television camera was turned on. As Mariner flew past Mars, travelling at a speed relative to the planet of almost 12,000mph (19,300km/h) and skimming to within 6,100 miles (9,800km) of its surface, the cameras imaged the landscape for about twenty-five minutes. Data was stored on a 330ft/100m-long tape spooled inside the spacecraft for later playback to Earth. Each 200 by 200 pixel image would take almost nine hours to download.

As Mariner 4 disappeared behind Mars, one more experiment remained. The radio signal was analyzed as the transmitting antenna vanished behind the planet, and the brief but measurable dimming of that signal gave a rough value of the density of the Martian atmosphere. After years of speculation, the atmospheric pressure of Mars, once thought to be perhaps that of a high mountaintop on Earth, turned out to be only about one thousandth that of our home planet. The first nail was pounded into the coffin of Percival Lowell's Martian civilization.

After Mariner passed Mars and headed off into deeper space, the slow playback of the pictures was initiated. Scientists at JPL received digital printouts of the data—just long rows of numeric values—from distant tracking stations around Earth via Teletype machines. Since these were numbers and not images, the data needed to be fed into the computers at the laboratory, a time-consuming process, to be converted into images. Unwilling to wait for the computers to create the first of twenty-two black-and-white photographs, a couple of the scientists grabbed the first printouts and hand-colored them with grease pencils. Each numeric value indicated a color, and it was like painting-by-numbers for PhDs! The resulting images were surprisingly close to the final computer-generated product, and still hang proudly at JPL to this day.

Eventually, the computers crunched the numbers and then printed the photos and the data sets. Early analysis of the results was slowly rolled out to the waiting media. The ghostly pictures showed a very different Mars to the one seen through the eyepieces of Earth's telescopes. The planet was a dry, rocky world, covered with more craters than anyone had imagined. Gone were the canals, gone were the oceans and continents. It appeared much like our own moon. The scientists were elated, the press engrossed, and the public, ultimately, torn between fascination with this technological marvel and shocked by the bleak landscape. With Venus and Mars reconnoitered, it was clear that neither planet was anything like a twin to Earth, nor was there a neighboring civilization. We were the lone sentient beings—at least within our own solar system.

But this was just the first outing. Mariner 4's twenty-two images were a discontinuous, low-resolution swathe taken diagonally to the surface. The mission, while a resounding success, had imaged only about one per cent of the planet, and happened to map only heavily cratered terrain. Details were few and much had to be inferred by comparing to terrestrial geology on Earth. This was the low point of Mars folklore—the realization that it was a completely dead wasteland, at least in terms of the small part of the surface that had been photographed. In the next decade, our understanding of the planet would head off in a new direction, ultimately as fascinating and inspiring as the one it replaced.

MARS IS NOT RED: MARSNIKS FAIL AND MARINERS SOAR

Heady from the success of Mariner 4, NASA made improvements to the Mariner design. A lightly modified version of Mariner 4 was sent to Venus as Mariner 5. The mission was a success, further boosting confidence in the basic Mariner concept. Work continued to upgrade the designs and subsystems for Mariners 6 and 7, due for another sprint past Mars in 1969.

AT THE SAME TIME, THE SOVIET UNION WAS NOT IDLE. Since the launch of Sputnik in 1957, the Soviets had enjoyed great success in their manned spaceflight program. Throughout the early 1960s they pulled off one space spectacular after another, launching the first man into orbit in 1961, the first woman into space in 1963, and completing the first spacewalk in 1965. However, their robotic space program had not fared as well.

LEFT: A fanciful USSR stamp commemorating the Mars 2 spacecraft as it rockets to the red planet. The probe made orbit in November 1971, transmitting data to Earth for over six months. The lander crashed.

ADVANTAGE USSR?

At first sight, the advantage in robotic spaceflight would seem to go to the Soviets. Along with robust guidance and navigation hardware developed for their own spacecraft and missile programs, they had far larger rockets that could carry much more massive probes into space. They took full advantage of this capacity, launching dozens of heavy and very ambitious missions to Venus and Mars from 1960 on. But appearances can be deceiving, and what went atop those large rockets could not match their leaner American cousins.

The USSR's early robotic efforts were promising enough: Luna 2, a small probe aimed at the moon, launched in 1959 and was the first spacecraft to leave Earth orbit. It did not, however, meet its primary objective of impacting the moon, sailing past it due to a timing error at launch. Luna 2 took that honor, smashing into the lunar surface, as intended, in September, 1959. Shortly thereafter, Luna 3 took the first grainy images of the previously unseen far side of the moon.

The next goal on the Soviets' robotic exploration agenda was Mars. While their numbering scheme was different from that of the

MARINER 6 & 7

MISSION TYPE: Mars flyby
LAUNCH DATE: Mariner 6: February 24, 1969,
Mariner 7: March 27, 1969
LAUNCH VEHICLE: Atlas-Centaur
ARRIVAL DATE: Mariner 6: July 30, 1969,
Mariner 7: August 4, 1969
MISSION TERMINATION: Mariner 6: July 31 1969,
Mariner 7: August 5, 1969 (flyby +1)
MISSION DURATION: 6 months
SPACECRAFT MASS: 910lb (412kg)

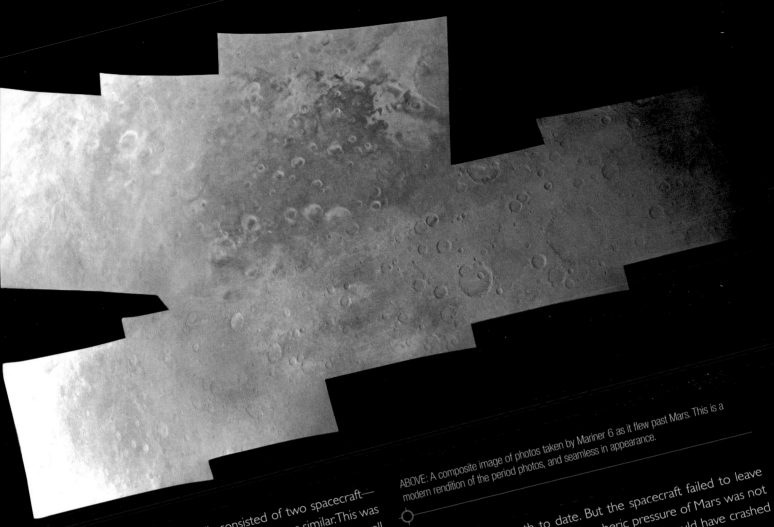

ABOVE: A composite image of photos taken by Mariner 6 as it flew past Mars. This is a modern rendition of the period photos, and seamless in appearance.

US—the Mars 1M mission actually consisted of two spacecraft—the purpose of launching their missions in pairs was similar. This was their largest robotic craft to date, weighing 1419lb (644kg), well over twice Mariner 4's mass. In late 1960 they tried twice to send these Mars probes aloft, but both launches failed before reaching their desired orbits, crashing back to Earth. The Western press dubbed them Marsniks. When a hurriedly prepared third launch attempt was made, at Soviet Premiere Khrushchev's insistence, the rocket stalled on the pad and technical crew was sent to investigate while the rocket was still fueled. This was against all safety guidelines, and the rocket exploded on the pad, killing the technicians.

In late 1962, the Soviets launched spacecraft called Mars 2MV-4 Number 3 and 2MV-4 Number 4, another set of twin spacecraft bound for the red planet. Once again, one failed before leaving Earth orbit. The second was successful leaving Earth, but communications with the probe were lost en route to Mars.

In a final attempt for glory in 1962, Mars 2MV-3 number 1 was launched. This was an entirely different spacecraft, despite the similar designation, which carried both a flyby vehicle and a primitive lander. The craft weighed almost 2000lb (900kg). It was an ambitious mission for the time, as no landing had been made

on any body off-Earth to date. But the spacecraft failed to leave Earth orbit. In any case, the atmospheric pressure of Mars was not correctly understood in 1962, and the lander would have crashed even if it had made the journey to Mars.

The next attempt took place two years later, a couple of days after the launch of Mariner 4 in November 1964. The mission was called Zond 2, but was essentially a clone of Mars 2MV-3, another large spacecraft with a descent module for a semi-soft landing attempt. The launch was successful, as was the insertion into a Martian trajectory. However, communication with the Zond 2 failed in May of 1965, and the craft sailed silently past Mars a month after Mariner 4's flyby. These three modes of failure—failed launches, failure to leave orbit, or failure en route—were becoming emblematic of the Soviet unmanned space program, and in particular in the case of attempts to reach Mars. Yet with incredible tenacity, relying more upon repeated attempts than high-tech, the Soviets prepared for another mission.

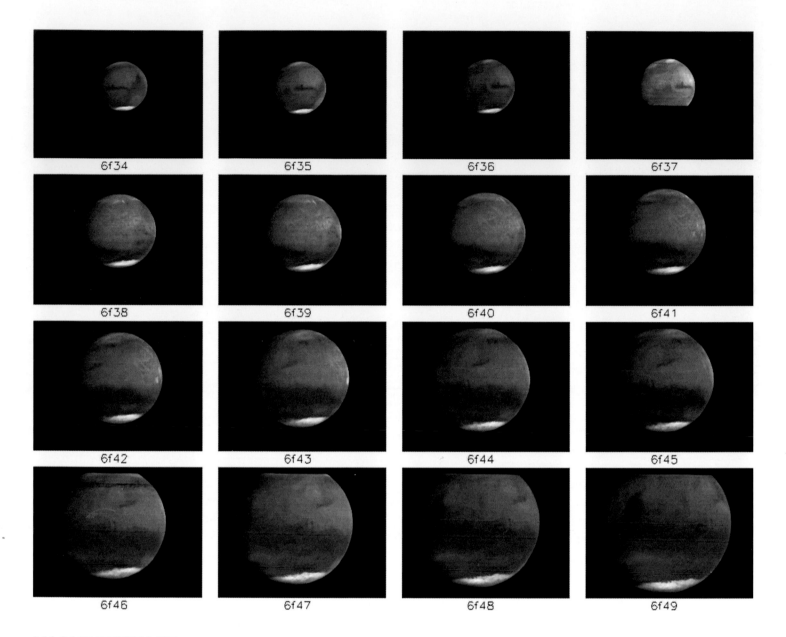

6f34	6f35	6f36	6f37
6f38	6f39	6f40	6f41
6f42	6f43	6f44	6f45
6f46	6f47	6f48	6f49

NASA'S RESPONSE

In the meantime, NASA was working hard on their next—and last—Mars flyby missions. Mariners 6 and 7 looked similar in overall appearance to Mariner 4, but were twice as heavy and once the solar panels were deployed, almost 20ft (6m) wide by 10ft (3m) high. Power was again provided by solar panels, supplying electricity to batteries inside the spacecraft. These were the largest and most massive unmanned spacecraft flown by the US to date.

Onboard experiments included an infrared spectrometer, an instrument for determining surface temperature of the planet, an ultraviolet spectrometer, an improved television camera, and another radio-occultation experiment similar to Mariner 4's. The radio was more powerful and data transmission rates were greatly

ABOVE: Sequential images taken as Mariner 6 approached Mars. Perhaps the most striking feature was the polar cap, seen at the bottom. Surface features can also be seen tracking from frame to frame.

improved. A tape recorder similar to that used on Mariner 4 was in place to playback the images once the spacecraft had passed Mars, but at much higher resolutions.

Mariner 6 launched on February 24, 1969, and Mariner 7 a month later on March 27. This time, the star-trackers worked perfectly and the cruise to Mars was largely uneventful—until a battery on Mariner 7 apparently leaked, causing the spacecraft's transmissions to fail days before the Mars flyby. Controllers switched to a backup

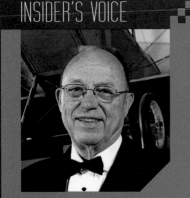

John Casani

Deputy Spacecraft
System Manager on the
Mariner Mars 67 Project

As Mariners 6 and 7 closed on Mars, everyone was glued to the readouts at JPL. Mariner 3 had been lost two years earlier, and Soviet failures had become almost predictable.

"All these failures got some to wonder about what might be going on out there. That's when the intriguing notion of The Great Galactic Ghoul came along … It came as Mariner 4 was approaching Mars. Since the Soviets had already lost two very close to Mars, people were speculating that there might be a cloud of meteorites surrounding Mars, or something about the Mars environment causing these problems. A writer working for *Time* magazine was asking me about these speculations and whether I thought there could be any truth to them. I said, 'No there's nothing in the Martian environment that should cause that. I can't explain why these other missions failed… maybe there's a space monster out there causing the problem.' So this writer immediately said 'a great galactic ghoul!' and that's where notion came from.

"A couple of years later we had a problem with Mariner 7 just as it was approaching Mars, evoking again the notion of the Great Galactic Ghoul. Some space artist produced a fanciful painting showing the Ghoul devouring the Mariner 7 spacecraft near Mars … Then people thought about all the missions the Russians lost at Mars, and the Ghoul took on a life of its own."

antenna, while engineers tried to come up with a fix for a mission profile they had flown only once before. Time was short.

Shortly after Mariner 6 sailed past Mars, proper radio contact was restored with Mariner 7. It was a close call, but the spacecraft seemed to be working fine again. Based in part on what had been learned through preliminary interpretation of Mariner 6's data, new commands were sent to Mariner 7, altering the mission parameters and selecting specific areas to be photographed. This was a strength of NASA's robotic spacecraft—reprogrammable computers. The computers used by the Soviet Union were generally less flexible and relied more on unchangeable command sequences programmed prior to launch.

Between them the spacecraft returned 210 images at much higher-resolution: 704 lines at 945 pixels each (compared to Mariner 4's 200x200). The best images were about 1000ft (305m) per pixel (Mariner 4's were on the order of 1 mile/1.6km or so per pixel), due partly to the close flyby trajectory, both craft passing just 2100 miles (3380km) above the surface of Mars. Both spacecraft had survived the flight to Mars, a rarity in early robotic missions.

THE PLAYING-FIELD LEVELS

The Soviet Union could only look on in envy. While they had their successes in Earth orbit with manned spacecraft, the technology of the US space program had long-since closed the early technology gap and surpassed the USSR's early successes in space. NASA was also more sophisticated in robotic command and control. Leonid Brezhnev, the Soviet leader, wanted to regain the lead once more, pressuring both the struggling Soviet manned lunar landing program (which ultimately failed), as well as other unmanned efforts aimed at Venus and Mars.

Topping the Soviet agenda for 1969 was the Mars 2M program, another twin mission. They were large and heavy, even by Soviet standards, tipping the scale at a whopping 10,700lb (4853kg) each. Each of the spacecraft had three separate cameras, a water vapor detector and an assortment of spectrometers. This was not a flyby mission—the twin spacecraft would orbit Mars.

The first Mars 2M launch was on March 27, 1969. After a promising liftoff, the third stage exploded and the spacecraft was lost. They tried again on April 2, but this time the rocket exploded only about 300ft (91m) above the ground. To add insult to injury, the caustic fuels released by the exploding Proton booster made the launch pad unusable for months, causing delays in other missions.

By now the Soviet space program was losing a tremendous amount of face, having failed to beat the Americans to lunar orbit, and soon to lose the race to land cosmonauts on the moon's surface. The early triumphs in both manned and unmanned spaceflight had faded into history, and NASA was surpassing Soviet space efforts on every front. And, while the USSR's missions were rarely announced prior to launch (US missions always were, except for military payloads), the word eventually got out. The worker's paradise was falling behind, and its failures at Mars would continue.

WET AND WILD: A SHOCKING VIEW FROM MARINER 9

After the loss of Mariner 8 in early 1971—which plunged into the Atlantic shortly after launching from Cape Canaveral—Mariner 9 was moved into position for a second try. This represented a new generation of Mariner spacecraft, again larger and heavier than its predecessors, with each new machine having more than twice the mass of Mariners 6 and 7 combined. This was partly due to the usual increase in scientific instrumentation, but also to a more sensational reason: these spacecraft were built to stay.

THE 1960S HAD MARKED THE LAST of the Mars flybys, with their quick over-the-shoulder look at the planet and a limited, opportunistic data return.

The added bulk of the new Mariners was due largely to an enormous fuel tank and rocket engine, intended to brake the spacecraft's hurtling trajectory once it reached Mars and drop into a stable orbit around that world. Almost half the flight mass was fuel. On May 30, Mariner 9 departed the Cape in smoke and flame, and within six months was approaching Mars. The critical rocket burn was just weeks ahead now, and would result in the craft braking and dropping into a comfortable mapping orbit—or, if it failed, unintentionally barreling past Mars, just like its predecessors. On November 14, the rocket ignited on schedule, burning for fifteen minutes and slowing the spacecraft sufficiently to allow it to be captured by Martian gravity. Mariner 9 became the first spacecraft in history to enter orbit around another planet... but it was not alone for long.

THE SPACE RACE ESCALATES

The Soviet Union had been doggedly pursuing similar goals in relation to Mars, but with little luck. However, seemingly each failure bred greater ambition. It was as if the roles of the catch-up game of the early space race had been reversed, and each new mission became more and more aggressive in the Soviets' attempts to catch up and leapfrog ahead of the Americans. And so, on December 2, the USSR's Mars 2 and Mars 3 joined Mariner 9 in orbit around the red planet. But these were not simply orbiters. The Soviet machines each carried a soft-lander and a small rover. It was an ambitious goal for the technology of 1971, and would have marked a spectacular set of achievements had the landers succeeded.

While the three spacecraft spawned of rival powers settled into their orbits, an unwelcome sight swam into view. Mariner 9 had returned unusually bland images as it approached Mars, and now the scientists soon verified what they had suspected since September: Mars—the entire planet—was enveloped in a massive, global dust storm. They could snap all the photographs they wanted, but the view would be only of the unbroken tops of the planet-girdling dust clouds. The disappointment was immense.

The engineers and controllers at JPL got busy, quickly designing new programming to be sent up to Mariner 9, delaying its mapping

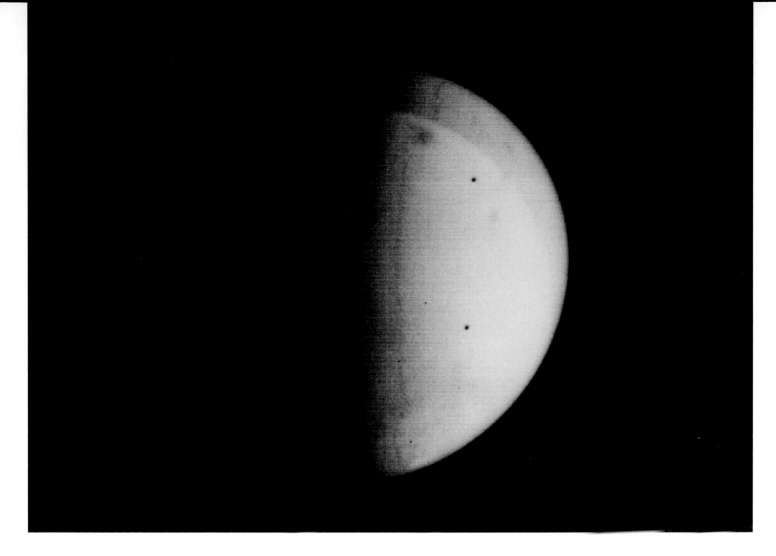

sequences. They put the imaging systems into only limited use, saving most of their capability until the storms cleared six weeks later. The USSR's Mars probes were not, however, reprogrammable, and went about their business, merrily snapping hundreds of images of blank cloud tops. Some useful data was returned, but not the scientific watershed they had hoped for. Mariner 9, meanwhile, bided its time for a clear shot of the world below.

And what of the Soviet landers? They were truly inspired designs for the era. About four and a half hours prior to arrival at Mars, each Soviet lander separated from the spacecraft, heading along a different, more aggressive trajectory that would plunge the machine into the Martian atmosphere. The combined mass of each orbiter and lander was well over 10,000lb (4535kg); five times that of Mariner 9. The landers were beasts of machines, weighing almost 2700lb (1225kg) fully fueled. They were, like many Soviet space machines, roughly spherical in design. As they approached Mars,

the landers would fire braking rockets, enter the atmosphere, and perform an automatic soft landing via a combination of rockets and parachutes. Once on the surface, four enclosing pedals would unfold, righting the probe and setting it into the proper position for operations.

There would have been one further surprise in store for the West had they succeeded: the Soviet landers each featured a rover, a 10lb (5kg) box about the size of a modern Xbox. These would have been deployed via a small robotic arm after landing, and once set on Martian soil would have scooted forward using a set of skids that would alternately lift and push the machine, not unlike the motions of a baby sea-turtle heading to the water after hatching. It was connected to the lander with a 50ft (15m) cable, and was somewhat autonomous, having bump-detectors in front. It was an ingenious, if rudimentary, design, but doomed to failure.

OPPOSITE: A Soviet-era stamp commemorating the missions of Mars 2 and Mars 3. The Mars 2 lander crashed, leaving its orbiter to survey the dust-enshrouded planet. The Mars 3 lander set down successfully, but failed within minutes. The orbiter suffered the same fate as Mars 2: images of a blank world.

ABOVE: Mariner 9's first views of Mars were stunning, but not in a good way. The planet was gripped by a global dust storm and the view was featureless. Able to reprogram the orbiter's computer, JPL waited out the storm and restarted their observing program after a few weeks.

BELOW: This image shows an erosional feature near Olympus Mons, the largest volcano on Mars. Along with many other eroded gullies on Mars, it sure looked like something created with flowing water.

SOVIET SETBACKS AND AMERICAN SUCCESSES

The Mars 2 lander hurtled into the Martian atmosphere, suffered a maneuvering malfunction, and crashed. The Mars 3 flew a proper trajectory to the surface using its autonomous navigation system, and settled with a clump onto the surface of Mars. The mechanical pedals rotated downward, orienting the lander upright, and the cycloramic camera was deployed. It was a spectacular moment when the first television pictures began to return to Earth…

And fifteen seconds later, it was all over. After transmitting a scant seventy lines of video data, the lander suffered a total failure. Speculation as to the cause ranged from an electrical discharge caused by the dust storm to faulty design. In any event, the ambitious missions of the Mars 2 and Mars 3 landers were at the end, even as the orbiters arrived overhead, imaging the featureless cloud-tops.

As the Soviet Mars 3 landing drama played out below, Mariner 9 continued to circle the planet, making scientific measurements and snapping the occasional photo to track the status of the dust storm. The scientists at JPL wondered how long a Martian dust storm could possibly last. Then, weeks later, strange objects appeared in a routine photograph taken to record the weather patterns. Three puzzling, circular spots appeared in one photo, and were soon joined by another. It was not until the positions of these objects were coordinated with existing maps that the scientists realized exactly what they were looking at. These were the tips of massive, ancient volcanoes poking through the cloud cover. One of them, Olympus Mons, was sixteen miles (twenty-five kilometers) high, or nearly treble the height of Mount Everest, making it the highest mountain in the solar system. The realization that these were enormous volcanoes poking through the dense dust clouds was a revelation, and another indication of the spectacular nature of the geographic features on this amazing world.

As the dust storm waned, mission planners had a chance to finalize Mariner 9's new objectives. The original plan had been for the absent Mariner 8 to map about seventy percent of the Martian surface during repeated orbital passes (which had been impossible during the previous flyby missions), and for Mariner 9 to study atmospheric and surface changes over time.

With the failure of the Mariner 8 launch, Mariner 9's mission plan now combined the most important objectives of both orbiters. It was a tall order, and the ability to reprogram Mariner 9's onboard computer was critical. However, by the time the planet's weather settled down enough to allow imaging to begin, JPL was ready.

For the next year, Mariner 9 orbited Mars with a minimum altitude of just over 1000 miles (1600km). From this vantage point, the images it sent home were breathtaking. The mission managed to map almost the entire surface, though with one spacecraft performing the work that two had been intended to do, the outcome was a slight compromise, with the polar regions imaged in lower resolution due to the probe's oblique orbit. The results, nonetheless, were spectacular.

The severity of the moon-like world revealed by Mariner 4's twenty-two pictures had been moderated somewhat by the better resolution and coverage of Mariners 6 and 7, but the views from Mariner 9—with its repeated passes over the planet—provided enough data to completely rewrite many assumptions about the

RIGHT: The USSR's Mars 2 and Mars 3 were identical orbiter/lander pairings. The spacecraft was large and heavy, not in small part because the Soviets used pressurized hulls to protect the delicate electronics. Neither mission was entirely successful.

BELOW: A detailed Mercator projection map created from Mariner 9 images. While not wholly accurate, such image assemblies were a valuable tool for planetary scientists engaged in unraveling Mars' geological secrets.

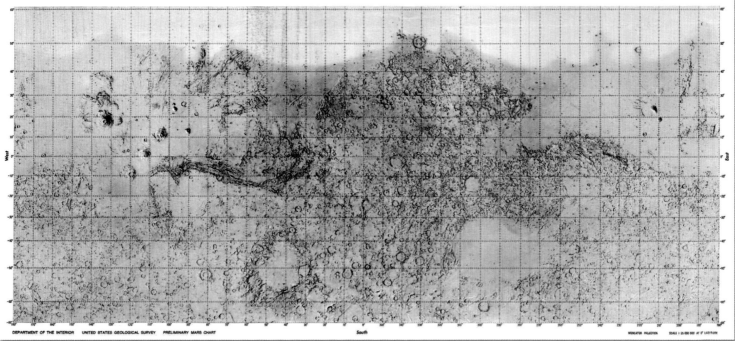

DEPARTMENT OF THE INTERIOR   UNITED STATES GEOLOGICAL SURVEY   PRELIMINARY MARS CHART   South   MERCATOR PROJECTION   SCALE 1:25 000 000 AT 0° LATITUDE

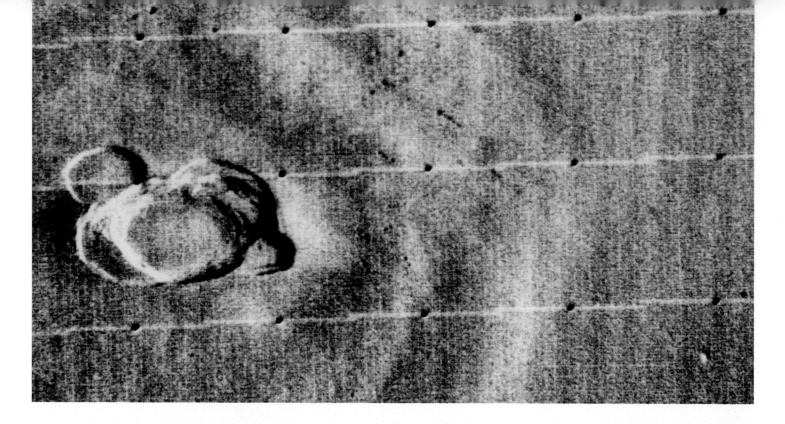

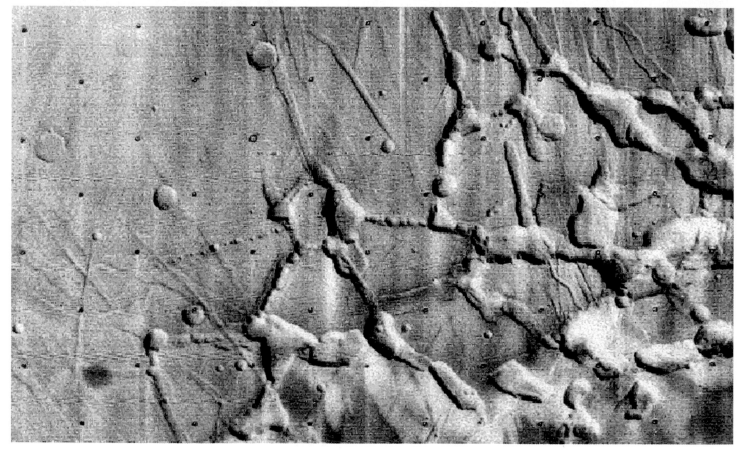

ABOVE TOP: The caldera of Olympus Mons, the gigantic Martian volcano that soars over 70,000ft (21,330m) above the surrounding terrain. This image was midway through processing, hence the TV transmission lines.

ABOVE: A view of Noctis Labyrinthus, or Labyrinth of the night, near the end of the Valles Marineris canyon. The bizarre structures still provoke questions regarding their formation, but are probably a result of erosional forces and collapse of the edges visible in the photo.

MARINER 9

MISSION TYPE: Mars orbiter

LAUNCH DATE: May 31, 1971

LAUNCH VEHICLE: Atlas-Centaur

ARRIVAL DATE: November 14, 1971

MISSION TERMINATION: October 27, 1972

MISSION DURATION: 349 days at Mars

SPACECRAFT MASS: 2200lb (997kg)

processes at work on the Martian surface. Unfolding below the cameras was a world swept with countless features that were clearly not impact-derived. Long, meandering channels—clearly formed by erosion from a once-active environment—were etched across the surface. There were areas that looked like river bends and deltas. It was not clear if these were water derived or the results of wind-driven erosion, but they surely looked like water-formed features as seen from Earth orbit. The geologists were elated—nothing is more thrilling to a planetary scientist than new questions to be answered.

Along with the global maps, the atmospheric composition of Mars was studied, including measurements of temperature, density, and pressure. An initial understanding of Martian weather patterns began to emerge, including the first look at seasonal changes. Vast amounts of data were returned, along with a total of 7,329 images. These would not only provide grist for years of study, but would also be useful for defining the mission parameters of the next set of Mars probes, scheduled to arrive in just four years. Mariner 9 had been a showstopper of a mission.

END OF AN ERA

After slightly less than a year, the fuel in Mariner 9's altitude control system was severely depleted and reorienting the spacecraft became impossible. With regret, but a deep sense of accomplishment, controllers turned the spacecraft off in October of 1972. Mariner 9 continues to orbit Mars, a silent monument to early planetary space exploration. It is expected to enter the Martian atmosphere in the early 2020s, and its debris will impact in a location as yet unknown. However, it's certain that the new, high-resolution cameras now orbiting the planet will be watching.

In just six years, Mars had evolved from a flickering smudge in telescopes to a bleak, cratered moonscape, and now to a planet eroded and abraded by some powerful weathering agent. In geological and meteorological terms it was a living, breathing ecosystem. But were these magnificent landscapes created by wind or water? And was the planet still active or were these all ghosts from a violent past? These were questions that would have to wait until a pair of vastly improved—and insanely ambitious—spacecraft would reach Mars in 1976… the magnificent Vikings.

INTO THE UNKNOWN:
THE MAGNIFICENT VIKINGS

They had been pondering this moment since the 1950s. The maps had been studied, the telescopic ones first, then the intricately prepared photomontages from the Mariners. The onboard instrument package had been debated endlessly. Mission parameters had been laid out, adjusted, and adjusted again. The landing site had been agonized over repeatedly. The hardware was built, tested, sterilized, and readied.

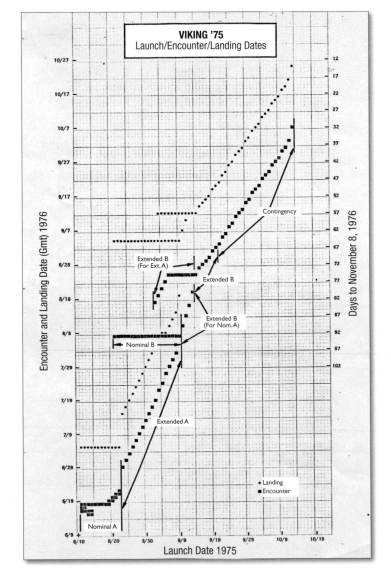

VIKING '75
Launch/Encounter/Landing Dates

Encounter and Landing Date (Gmt) 1976

Days to November 8, 1976

Contingency

Extended B
(For Ext. A)

Extended B

Extended B
(For Nom. A)

Nominal B

Extended A

● Landing
■ Encounter

Nominal A

Launch Date 1975

THEN, IN AUGUST AND SEPTEMBER OF 1975, two Titan III rockets left Cape Canaveral to deliver their heavy payloads into earth orbit, after which the upper stages headed off to Mars.

In June, with braking rockets firing, the first of the two Viking spacecraft—twin explorers with orbiters and landers traveling in tandem—settled into Martian orbit. From there, via images from the orbiters' new, enhanced cameras, scientists at JPL contemplated the jumbled, chaotic surface of the planet below for a whole month. This was their last chance to alter plans for the best landing site.

Then, just before 1am in Pasadena, California, on July 20, 1976, the Viking 1 lander detached from its orbiter, fired braking rockets and headed toward the surface of Mars. It was, by today's standards, a large, heavy, and not very intelligent machine headed into an information abyss. Little was known about the Martian surface even then, and landing successfully—meaning, not setting down on a large lander-destroying rock or straddling a crater rim—would

Viking missions to Mars

Expanding human knowledge **USA 15c**

mostly be a matter of sheer luck. Today, many at JPL refer to the Vikings, with great affection, as BDLs… "Big Dumb Landers."

For the next three hours, Viking 1 would be operating primarily on its own. It would either greet the Martian surface at a leisurely pace, or with a terminal, lander-shattering impact. And, due to the long radio delay between Earth and Mars, nobody at JPL would know the outcome one way or the other for about nineteen minutes. Nervous flight controllers fidgeted at their consoles, awaiting the all-critical first transmission from the surface…

A NEW DEPARTURE

The Viking missions were a culmination of well over a decade of planning and intensive preparation. Before the first Mariners scooted past Mars, people within NASA and academia at large were already planning a robotic landing mission to explore the planet's surface. Looking at Mars from Mariner 9's orbit was a fine way to do science, but orbiters alone could never tell the whole story. Nothing could match the data to be gleaned from sending a combined orbiter/lander, and the Vikings were to be the ultimate expression of human curiosity about Mars.

Between the data from the surface and the continued observations from orbit, our understanding of Martian geology and weather processes would be transformed. New high-resolution orbiting cameras would return images of unparalleled precision, and this time they would be in color. The landers were massive, packed with extensive instrumentation to explore everything from soil composition to Martian weather and seismic activity. But topping the agenda was a daring quest for a human connection with the red planet: the search for life beyond Earth.

From the first years of planning the mission, the notion of searching for life on Mars had been part of the agenda, and as the mission came together, this goal assumed front-row status, at least so far as the press were concerned. There had been much debate, primarily over whether to concentrate the surface science investigations on a search for water, or jumping ahead to look for microbial life in the soil. In what may now appear to be a bit of scientific hubris, but based on sound reasoning at the time, it was decided to seek signs of microbial activity in the Martian soil. This would turn out to be a long-odds bet that was not rewarded with a clear, black-and-white answer.

While the orbiter and lander were being designed and built, the search was begun for a landing site. Data from Mariners 6 and 7

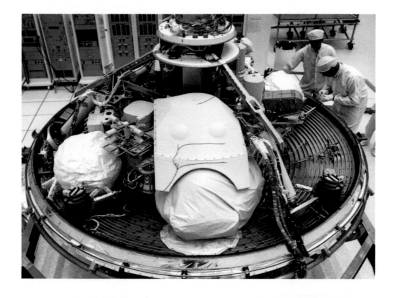

OPPOSITE LEFT: A timeline of important mission events – planning the Viking missions continued until launch with every process documented.

OPPOSITE RIGHT: In this commemorative US stamp, a Viking lander is "on top of the world" as a tiny Mars awaits sampling.

ABOVE TOP: Technicians with the Viking lander. Underneath the assembly is the 11.5ft/3.5m-diameter heat shield. It was the first time NASA used such a structure in any atmosphere other than Earth's.

ABOVE: A technician checks the function of a Viking lander's soil sampling scoop. The scoop was affixed to a robotic arm made of spring steel that rolled-up on a reel to retract.

Norman Horowitz

Professor of Biology, Caltech

Horowitz oversaw the development of the life science experiments for the Viking mission.

"The idea of life on Mars has been around for three hundred years. And here was the first time we had the ability to test it. Mariner 9 found an objective argument for flying to Mars, because Mariner 9 saw that Mars once had water on it. There are dry streambeds, obviously cut by water. All the geologists agree that they're water cut; there was water on Mars at one time. And you could say that, if there was water on Mars, then there may have been an origin of life, and that life may still be surviving. Now, Mariner 9 was an orbiter, and it orbited Mars in 1971; up to that point, up to the time Mariner 9 took its photographs, I would have said the a priori probability of life on Mars was close to zero.

"If we are the only inhabited planet in the solar system, and there's only one form of life on Earth—I mean, when you look at the composition of living creatures and see that they all have the same genetic system and they all operate on DNA and proteins composed of the same amino acids with the same genetic code, it's clear there's only one form of life—then we're all related. The origin of life may have happened only once, and it happened here and no place else in the solar system. Or if it happened elsewhere, it didn't survive. I think this is a conclusion of really cosmic importance. If people become aware of this, then maybe they'll be less inclined to destroy the planet."

was informative but of little comfort to those who were charged with delivering the machine safely to the surface. The images from these spacecraft covered just a small percentage of the planet and were of sufficient resolution to avoid landing inside a small city, but that was about it. There was simply not enough information to make a data-driven decision. Then, new data came in from Mariner 9 in 1971. It offered better resolution and improved coverage, but could still not identify an object smaller than about 0.7 miles (1.1km) in diameter. Landing zone selection would be as much about intuition as observation.

DIVERGENT OPINIONS

Added to these safety concerns were the tensions between members of the science team regarding a preferred landing location. While they all wanted good science, their goals pulled in various different directions. The meteorologists wanted to select an area that was flat and clear for miles all around, perfect for getting clean weather data but anathema to the geologists, who wanted interesting surface features nearby that they could photograph. Also, a good-sized crater nearby might have thrown-up some material from beneath the surface that they could sample with the robotic arm. The biology team, working from the same low-resolution orbital imagery, felt that placing one of the landers in a light-toned area, and the second in a dark-toned area, might give them a

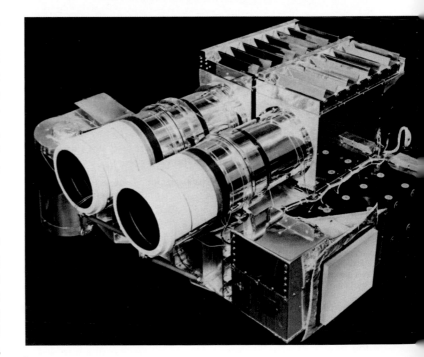

ABOVE: The stereo Viking orbiter cameras were hi-resolution "slow-scan" TV cameras with telescopic lenses. Six different colored filters could be rotated into the optical path to restrict certain bandwidths of light.

OPPOSITE: A composite image of Mars as seen by Viking 1. Of particular interest are the north polar cap in top center, and the sprawling Valles Marineris crossing the bottom half of the planet.

better chance of finding life—it was still not clear to them what processes were involved in the differing coloration. The chemists wanted a low-lying area that might have denser atmospheric pressure, be warmer and, possibly, wet.

The meetings continued into 1970. While they would work with image data from Mariners 6, 7, and 9, the final decision on a landing site would be confirmed only after they had images back from the new Viking's own new orbiter cameras. And in the midst of these debates, NASA headquarters dropped a bomb: the agency wanted the Viking team to cut costs, and those same improved orbiter cameras were on the chopping block. The choices before them were to use older cameras of the Mariner 9 design, simplify the existing camera, or eliminate cameras altogether. It was the

Mariner 4 debate all over again, but amplified to mission-critical proportions. Without the improved cameras now being designed for the Viking orbiters, landing site selection would be almost pure guesswork.

The scientists and engineers at JPL pushed back, and the arguments raged long and, at times, loud. In the end, the improved orbiter camera design was retained, offering resolutions far better than those of the Mariner 9 units. There was simply too much to be lost by eliminating or simplifying them, both in science and in landing zone selection (by the time the second Viking orbiter concluded operations in 1980, nine thousand images had been radioed back, and many led to unique discoveries that could simply not have been obtained with less powerful imaging).

LEFT: Twenty-two separate images were combined to create this spectacular shot of the top of Olympus Mons. This is the caldera, or volcanic crater. Up to seven craters are blended into one giant depression atop Olympus Mons, and the caldera floor is as much as 2.5 miles (4km) below the crater rim.

OPPOSITE: A Viking lander, encased within its protective aeroshell, undergoes sterilization at JPL's heating facility. Ten percent of Viking's $1 billion budget was spent towards the sterilization and cleaning of the landers.

ADVANCED TECHNOLOGY

Two instruments had been defined as essential early in the selection process: a gas chromatograph and a mass spectrometer. Both would be able to detect the chemical elements in very small amounts of gas exuded by heated soil samples. This was the first deployment of such devices to space; both have returned to Mars in highly evolved forms since, as recently as the Curiosity mission that began in 2012.

At the core of the arguments over the remaining experiment options was the notion of whether or not any microbes on the planet would metabolize nutrients similarly to Earth-based organisms. In a final compromise, the life science package included experiments that should detect microbial life in either case—whether it ingested and metabolized on-board nutrients (which would be squirted into the soil sample) or not. As with most compromises, it thrilled neither side, but was sufficiently placatory to both to move things forward. Three more experiments were chosen with an eye toward sensing microbes in the soil.

Meteorology instruments, a seismometer, another spectrometer (one that used X-rays), and the camera package completed the science package on the landers. These instruments would measure daily weather patterns, look for Marsquakes—which could help to understand the interior structure of the planet—and, via the X-ray spectrometer, determine the general mineralogical composition of the Martian soil.

When the Viking spacecraft were completed, they needed to be tested and prepared for the difficult journey to Mars. Other than

But improved imagery from the orbiters was still a couple of years away in the future, and the work to refine the mission, and make a decision where to drop the landers, continued.

A core concern was the life science experiment. Weather and geology were important, but the life science experiments, which dominated the lander's agenda, were critical. The question of how best to detect life on another world—with the limited technological and financial resources available to JPL, as the Apollo program was sweeping most of the money off the table—became its own debate between the life science scientists.

VIKING 1 & 2

MISSION TYPE: Mars orbiter, Mars lander
LAUNCH DATE: Viking 1: August 20, 1975,
Viking 2: September 9, 1975
LAUNCH VEHICLE: Titan III-Centaur
ARRIVAL DATE: Viking 1: June 19, 1976, Viking 2: August 7, 1976;
Viking 1 landing July 20, 1976, Viking 2 landing: September 3, 1976.
MISSION TERMINATION: Viking 1 orbiter: August 17, 1980,
Viking 1 lander: November 13, 1982, Viking 2 orbiter: July 25, 1978, Viking
2 lander: April 11, 1980
MISSION DURATION: 2307 days; 6 years,
4 months on Mars (Viking 1 lander)
SPACECRAFT MASS: Orbiter: 1947lb (883kg), Lander: 1261lb (572kg)

the Soviets' Mars 3 mission in 1971, which ended only seconds after touchdown (and the uncontrolled crash of its twin, Mars 2), nobody had landed a machine on another planet. NASA was very concerned about protecting the Martian environment. Stray germs from earth, carried aboard the spacecraft, could not only promote false detections of life, but could potentially contaminate Mars for evermore. While many scientists regarded this as being a very unlikely scenario, NASA was taking no chances. The sterilization procedures used on the Viking landers accounted for one tenth of the mission's overall cost, and these protocols are still the gold

standard used for planetary landing missions today. The issue of whether or not this effort is worth the cost is still hotly debated within the planetary exploration community, but NASA's opinion continues to be that it is better to be safe than sorry.

The landers were scrubbed clean again and again during assembly and preparation, then cleansed a final time before being taken off to JPL's large sterilization chambers. Here, they were baked for thirty hours and thirty minutes at 233°F (112°C) before being sealed in the protective aeroshell (the cover inside which the landers would ride to the surface). Today's landers and rovers are not subjected to this level of sterilization, due to both cost and the dangers to the sensitive electronics within. However, any future missions that are slated to land near a source of liquid water, and therefore possibly life, will have to revisit the Viking protocols.

Once assembled, clean and sterile, the entire package was bundled off to the Kennedy Space Center in Florida. There the machines were rechecked and stacked atop Titan III rockets, which were NASA's most powerful boosters at the time.

On August 20, 1975, Viking 1 rocketed out of the Cape. Viking 2 followed it on September 9. After reaching Mars a half-year later, the two spacecraft orbited for a month while the science teams looked over the terrain below. The image resolution was still too low to predict exactly where the best site would be, but it was enough of an improvement over Mariner 9 to cause several last-minute adjustments. Seven years to the day after the landing of Apollo 11, the Viking 1 lander headed for the surface of Mars. The events of the next few days would electrify the world.

PINK SKY, RED SAND

The 7×10ft (2×3m) lander plummeted into the upper atmosphere at over 10,000mph (16,000km/h) with braking rockets firing. It had been well over three hours since the Viking had separated from the orbiter and begun its descent to Mars. The lander was steering its way through this increasingly violent phase of its flight with its tiny 18kb computer by firing twelve thrusters arrayed around its protective aeroshell. The heat shield was doing its job, insulating the lander from the 2,700°F (1,482°C) heat outside.

AT JPL MISSION CONTROL, technicians looked at their screens with increasing intensity, reviewing data sent nineteen minutes earlier. It took that long for the signal to reach earth, so it was old news. All they could do was watch and wait... Viking was on its own.

At an altitude of 17 miles (27km), after falling hundreds of miles from orbit, Viking adjusted its trajectory to a steep glide, scrubbing excess speed. By actually generating lift via its heat shield, the spacecraft could continue to slow from its high-speed entry into the atmosphere by gliding horizontally for many miles. At 19,000ft (6000m), at just under 1000mph (1600km/h), the massive parachute was fired out of mortars mounted atop the aeroshell. Despite the concerns of those involved in the manufacture and testing of the 52-ft (16-m) cloth dome, it did not rip or tear, but inflated to full volume, slowing the craft. The heat shield was jettisoned, and three landing engines arrayed around the bottom of the lander began firing. A radar altimeter sent range-to-ground data to the computer, relaying the lander's altitude. What it could not determine was the roughness of the surface below. That was going to be a matter of dumb luck.

THE MOMENT OF TRUTH

As this mechanical and electronic ballet was performed, the lander was already pursuing scientific investigations. The pressure and composition of the Martian atmosphere had been under continuous analysis since shortly after atmospheric entry. Nothing is wasted in spaceflight; not even the data to be gleaned from a searing, intricate first fall through the Martian atmosphere.

The parachute and top cover were released forty-five seconds before touchdown, and Viking 1 was now headed straight down—any horizontal motion would snap the landing legs like brittle twigs. It slowed further, finally touching the surface at 6mph (10km/h), about the speed of a brisk walk. The engines immediately shut down, and all was silent. After a 440-million mile (708-million kilometer) journey, Viking was finally on Mars.

Back on Earth, the mission controllers jumped and clapped

LEFT: The NASA JPL mission patch worn during Project Viking.

RIGHT: Viking program engineers watch the landing of Viking 1, July 20, 1976.

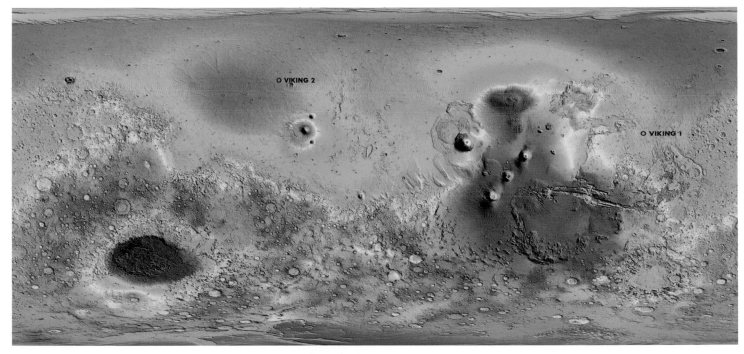

ABOVE TOP: A 1976 image sent back by Viking 1. The thin Martian atmosphere is clear at the planet's edge.

ABOVE: Viking's landing sites were spaced well apart, almost on opposite sides of the planet.

RIGHT: Olympus Mons as seen from a Viking orbiter. The caldera (craters) at the top are at their widest 50 miles (80km) across.

with joy, some forgetting that their bulky headphones tethered them to their consoles. Many in the room had given the landing a fifty-fifty chance of success, though few had mentioned it out loud. That time of nervousness was now behind them, and the lander, now slowly cooling on the Martian surface, would soon begin operations.

Viking had set down well within the planned landing area, 22.8 degrees north of the Martian equator, in a region called Chryse Planitia, Greek for the Golden Plain. It was a region chosen as much for its apparent safety as for scientific goals—while appearing to be a depression that seemed to have had a number of sources of ancient water aimed into it, it also looked reasonably smooth and not too densely cratered. What kind of wind-and-water carried riches might await the first machine to land successfully on Mars?

The Viking landers were designed for a ninety-day mission. Each was capable of carrying out the basic functions of that mission autonomously if necessary, a precaution built into its primitive programming by the designers. However, with the Viking orbiters continuing their regular arcs overhead, that contingency was unnecessary, since the orbiters would act as relay stations, while continuing their own missions to observe and map the surface from space.

The first item of business was to test the lander's twin cameras and relay home an image for the engineers. International media were anxiously awaiting the first visual transmission from the surface of Mars. With no internet to allow instant updates on the mission, only those in a facility with the ability to relay NASA's direct link would see the images in real-time. The cameras were a new type of design—rather than being equipped with a single video pickup and lens such as used on previous spacecraft, the Viking cameras looked like a vertically mounted coffee can with a slit cut into the side. Inside was a mirror on a pivot that reflected a small percentage of the desired image into a lens directly below, and past that, a cluster of photodiodes. The mirror would tip, scanning the terrain vertically, then rotate slightly and repeat the procedure, dozens of times.

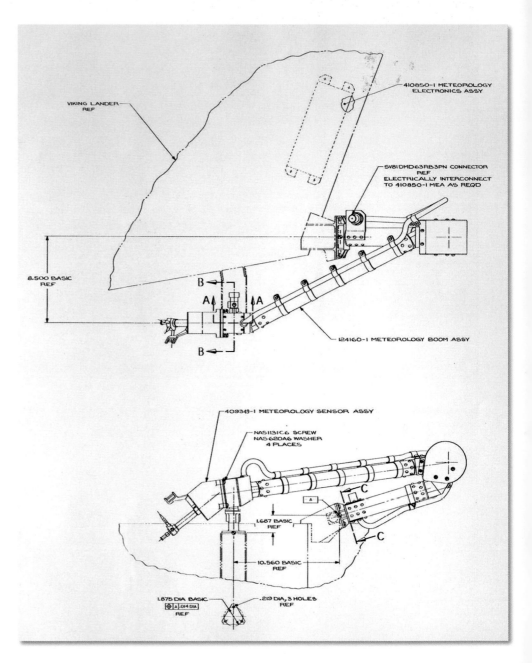

Dark and light values were converted into data to be sent to earth. It was a slow, painstaking way to gather an image, but allowed for unprecedented resolution, great color rendition, and even three-dimensional photography.

A PICTURE PAINTS A THOUSAND WORDS...

Mission planners wanted to capture two images right away. The orbiter would pass over the horizon in about fifteen minutes, and this might be all the data that they would get if the lander failed before the next pass. The first image came down to Earth after the long radio delay, assembling one vertical strip at a time, slowly

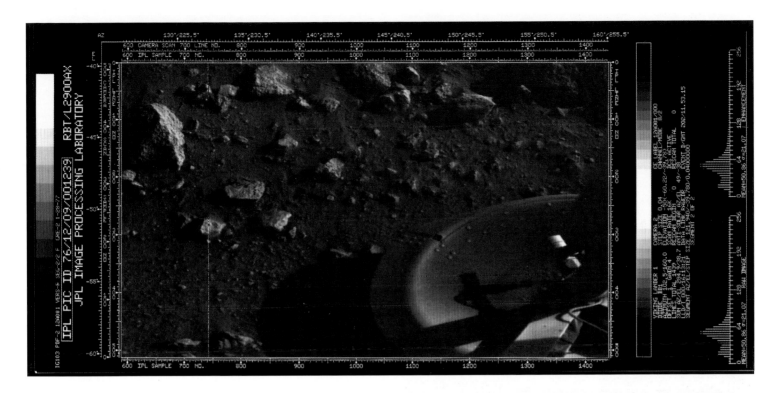

OPPOSITE: A schematic of the meteorology boom, which deployed by swinging up shortly after landing.

ABOVE TOP: The first image from the surface of Mars: one of Viking 1's footpads. While the journalists wanted a color panorama, the engineers wanted to make sure the lander had settled level and onto a firm footing.

ABOVE: A view of Viking 2's landing site: Utopia Planitia. The indications surrounding this image and that above is engineering data involved with the imaging process.

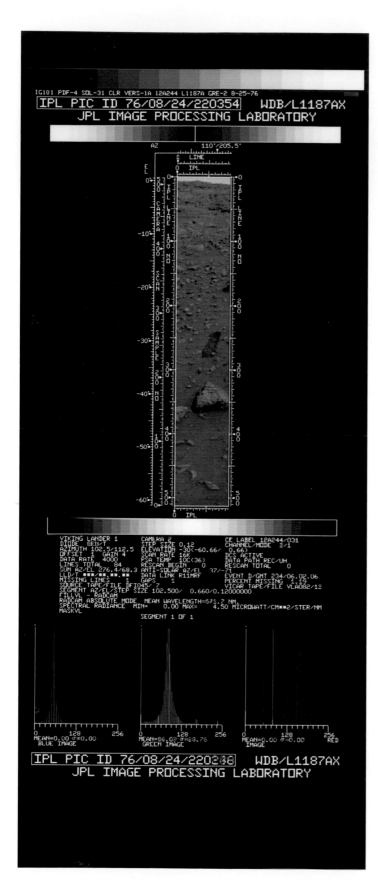

building from left to right. This first, historic shot was in black and white. Color images would be taken later, but they took much longer to complete. The black-and-white strips built up, one at a time, until the entire sixty-degree image was finished. The first image from the surface of Mars was… one of Viking's footpads planted on the rocky soil.

The assembled press grumbled—where was the spectacular shot of a distant, forbidding Martian horizon that all those tax dollars were supposed to have paid for? Indeed, that kind of first picture had been discussed, but mission planners had intelligently determined that ascertaining that the lander was securely settled onto level ground had to be the first priority. The panoramic vista would have to wait, and so would the reporters.

Immediately, the cameras were repositioned to shoot the first picture of the horizon, and it was spectacular: a vista of rocks and soil, rough and hummocky, stretching off into the hazy distance. This, too, was shot in black and white. A color image was received later, and once properly processed showed a salmon-pink sky and brick-red, highly oxidized soil and rocks. It was a sensation.

As checkouts were completed, a series of operations to release various mechanical parts of the lander was initiated. Small explosive charges were fired. A metal arm, the meteorology boom, swung up and locked into position. Soon the lander was sending home data on wind speed, atmospheric pressure, humidity, and temperature. Weather reports from Mars would be a part of the daily news for some time: average temperature: -67°F (-55°C); temperature range: -22 to -139°F (-30 to -95°C); atmospheric pressure: .09 PSI. Wind

speed and humidity were dutifully recorded for months, then years.

When it came time to release the seismometer, there was a hitch. The delicate sensing device was designed to be held firm during the violent rides into orbit and down to Mars. But a retaining pin that was held in place by a piece of wire that was to be severed electrically had not disengaged. Viking 1 would detect no Marsquakes for the duration of its mission. NASA could only hope that the second lander did not suffer the same fate.

There was more drama in store. The lander's robotic sampling arm would not extend properly. Since this was the only way to pick up soil samples and deliver them to the extensive laboratory instruments inside the lander, it became a primary concern. A locking pin that retained a protective shroud was stuck, partially blocking the arm. JPL engineers huddled with the team that built the lander at Martin Marietta, and soon devised a plan to re-extend and then partially retract the arm, which should allow the pin to release. Within a few days, revised software commands were sent to the lander, and upon their execution the pin fell free, allowing the arm to become fully mobile. Viking would be

able to fulfill its primary function—gathering and analyzing soil samples—after all.

On July 28, just over a week after landing, the arm was fully extended. It dug a couple of test trenches, which revealed much of the soil characteristics within the lander's "grab radius," or area that that sampling arm could reach from the static lander. Once they understood that they were indeed dealing with soil that was mechanically similar to that found in Earth's deserts, a small sample of dirt was scooped, slowly swung over the intake funnels, and sprinkled into the analysis instruments. The assembled controllers, engineers, and science teams settled in for the wait. While just a part of the science, this was in effect the holy grail of the mission: would there be a detection of life on Mars?

However, few things in space exploration are simple, and often any questions answered end up prompting even more. In this case, the data did not lend itself to a simple, straightforward conclusion as hoped. The results of the Viking life science experiments fell into this category, and ultimately provoked debate and controversy that have endured to this day.

OPPOSITE LEFT: A trench made by Viking 1's sample arm. This is indicative of images made from one or a few single vertical scans of the camera. Engineering data surrounds the color picture.

OPPOSITE RIGHT: Before and after shot of the sample scoop and the trench it created. The reach of the arm was a small arc on one side of the lander.

ABOVE LEFT: Viking 1 and the rock called Big Joe at Chryse Planitia.

ABOVE RIGHT: Viking 1 took this picture 15 minutes before sunset on August 21, 1976, one month after landing.

IS IT LIFE?

Once the soil sample had been safely delivered to the instruments, the science teams got busy starting their experiments. Some went faster than others—the biological processes needed for two of them took time to generate data. The gas chromatograph—mass spectrometer (GCMS) looked at raw, untreated soil that was heated to varying degrees and the resulting gas released was analyzed.

THE ANALYSIS SHOWED NO ORGANIC CARBON. The instrument was accurate to a few parts per billion, and this was not encouraging news regarding the possibility of life. But there were three more experiments in the life science package.

Each of the other experiments was predicated on the notion that all microbes release something—a waste byproduct—in the process of gathering energy from food. Little was known in the 1960s and early 1970s of the many types of extremophiles that have been discovered since then, life forms that can exist in extreme environments, such as inside Antarctic rocks or near dark, scalding deep ocean vents. At the time traditional microbial metabolism seemed like a good way to test the Martian soil for living organisms. It could have worked—had there been any microbes present, and had the Martian soil chemistry been a bit different.

The gas-exchange experiment (GEX) took a soil sample and pumped out the Martian atmosphere from the test chamber,

replacing it with helium. A broth of nutrients was squirted into the soil, and purified water was added. Throughout the multi-day process, samples of the air inside the chamber were examined, with a particular emphasis on searching for oxygen, nitrogen, hydrogen, and methane. The hope was that any microorganisms in the sample would metabolize the nutrients and create measureable byproducts. After a number of runs, sadly the results were conclusively negative.

THE EXPERIMENTS CONTINUE

The pyrolitic release (PR) experiment (also known as the carbon assimilation experiment) added water, artificial light, and a replicated Martian atmosphere with a radioactive component—carbon 14—to its sample. After a number of days, the air inside the test chamber was pumped out, and the remaining soil sample heated to 1200 °F (646°C). The gases resulting from the scorching heat were then examined. The idea was that any carbon 14 that had been metabolized by microorganisms would have been left behind when the air was pumped out, and spotted when any microorganisms remaining were burnt. The results were, once again, negative—there were absolutely no indications of Martian life.

The labeled release experiment, the third in the life science package, was the last hope for a positive indication of exobiology. Martian soil was injected with nutrients (much as had been done

LEFT: The Viking lander's sample scoop was able to dig trenches, collect sand and soil, and even nudge rocks to expose the fresher surface below. To deliver the sample to the instruments inside the lander, it sifted the soil down into the collection funnel.

OPPOSITE: This display of a Viking lander shows the full extent of the sampler arm. The arm itself is made of two pieces of spring steel, rather like two metal tape measures facing each other. When it was unwound from a spindle, the two halves popped into a tubular shape.

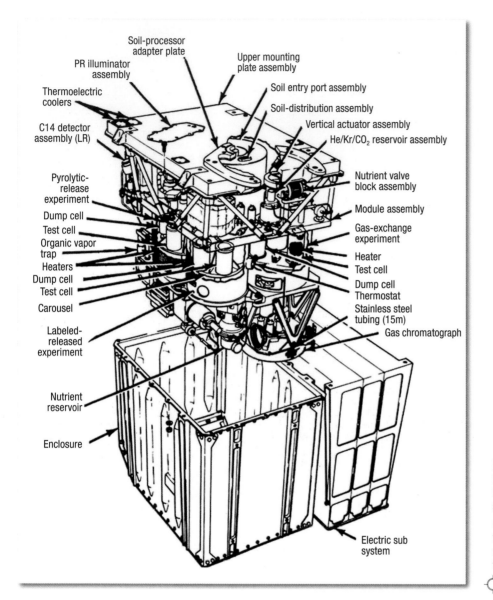

Soil-processor
adapter plate

PR illuminator
assembly

Upper mounting
plate assembly

Thermoelectric
coolers

Soil entry port assembly

Soil-distribution assembly

C14 detector
assembly (LR)

Vertical actuator assembly

He/Kr/CO$_2$ reservoir assembly

Pyrolytic-
release
experiment

Nutrient valve
block assembly

Dump cell

Module assembly

Test cell

Gas-exchange
experiment

Organic vapor
trap

Heaters

Heater

Dump cell

Test cell

Test cell

Dump cell

Carousel

Thermostat

Labeled-
released
experiment

Stainless steel
tubing (15m)

Gas chromatograph

Nutrient
reservoir

Enclosure

Electric sub
system

LEFT: The Viking lander biological experiment carried four instruments, each of which took a slightly different approach in an effort to identify microbes in the soil. While the evidence seems to show a negative result, some scientists are not so sure.

OPPOSITE: Day-to-day operations were carefully planned and tracked with handwritten memos. This 1977 form, filled out nine months after the landings, is a request to throw a switch onboard the lander.

in the GEX experiment), which included carbon 14. The air in the chamber was continuously monitored, the scientists looking for byproducts of metabolization of the nutrients by any living things in the soil.

In short order, the measurements indicated radioactive carbon dioxide (CO_2). Something in the soil seemed to be reacting to the nutrients, and this could have represented microorganisms digesting the radioactive broth. The graph continued to climb, and there were a lot of cautious smiles at JPL. A second dose of nutrients was added after a few days, and once again the graph took off. It was an impressive result, but a bit odd… the readings varied more quickly than one would expect from bacterial metabolism. However, this was Mars, after all, so anything seemed possible.

Finally, after about a week, the chamber was purged by heating, and then the experiment repeated on the same (now presumably sterilized) sample. As hoped, no metabolic gases formed, leading to the possible conclusion that whatever had been digesting the broth was killed by heating.

The lead scientist on the experiment, Gilbert Levin, was thrilled. To him, it seemed as if they had found life on Mars. Not so fast said others, primary among them Norman Horowitz, the scientist in charge of the PR experiment. What was found could have also been the result of oxidizing soil chemistry reacting to the water in the nutrient broth.

These results were debated at some length, but soon the majority of the scientists were forced to set the issue aside to concentrate on other aspects of the mission. There was much to do and a second lander to set down on Mars. But Levin was insistent: the results looked compelling to him. Horowitz was just as adamant in his opposition: it seemed to him as though these readings were the result of something highly oxidizing in the soil—possibly the caustic chemical, perchlorate. The difference in interpretation has lasted beyond Horowitz's death in 2005. Levin continues to advocate his point of view: yes, there are multiple interpretations, but one of them—a likely one in his mind—is that the results of the experiment offered a positive indication of life on Mars. Recent research has demonstrated that under certain conditions, microbial life existing in the presence of perchlorate can produce results similar to what was observed in 1976. Levin has campaigned hard for the inclusion of a new life science experiment on Mars landing missions since, but NASA has followed other priorities. The presence of perchlorate on Mars was later verified by the Phoenix lander.

VIKING 2 TAKES UP THE GAUNTLET

On September 3, it was Viking 2's turn to land on Mars. As with its predecessor, images from the orbiters had been carefully scrutinized and the potential landing site hotly debated. An area named Cydonia had been initially selected, but as the mission scientists agonized over the new images coming in, it looked worse and worse—too many potentially crippling obstacles were seen. Of course, with the smallest visible details about the size of the Rose Bowl or Wembley Stadium, it was impossible to tell what was really there; it was a matter of reading the trends in the terrain. As before, the scientists could only make educated guesses, finding the sweetspot between safety on the one hand and scientific return on the other. Again, radar data was included in an attempt to find a survivable landing zone.

The team voted and a decision was reached: Viking 2 was programmed to land in Utopia Planitia (the "Nowhere Plain"), 4,800 miles (7,725km) distant from Chryse Planitia but still in the northern hemisphere of Mars. Viking 2 made it down safely, but not without drama. Moments after separating from the orbiter, the lander experienced a power loss to its guidance gyros and went into a tumble, losing radio contact with Earth. Within minutes, a backup guidance unit took over, stabilizing the lander, and it headed down to Mars without intervention from the ground. As before, the controllers were nothing more than passive observers,

as the machine made its own decisions as it headed towards a successful touchdown.

Procedures and results were close to those of Viking 1. The life science package returned similar readings. Even the labeled release experiment produced results akin to those from Viking 1. But this did little to alter the opinions of the scientists. Environmental conditions were slightly different, but not outside of their expectations after two weeks of analyzing Viking 1 data. To the relief of the geologists, the seismometer on Viking 2 deployed properly this time and seismic data was available for the mission's duration.

Over the course of the next few years, the Viking orbiters and landers worked in concert to analyze Mars and its environs from both orbit and on the ground. Numerous soil samples indicated

iron-rich, highly oxidized soil of volcanic origins. The second lander detected seismic activity—known as "Marsquakes"—and a more complete understanding of the atmosphere and weather was gradually assembled, using readings from both sides of the planet.

And then there were the images. Each lander performed tirelessly, returning thousands of images, and the orbiters did the same. In all, the mission returned almost 50,000 images, many in color and some in 3D pairs. It was a spectacular return on the roughly $1 billion investment (in 1970s dollars) that had been made.

THE VIKING ENDGAME

All the components of the Viking mission outlived their ninety-day life expectation, but over the next few years, one by one the machines fell silent. Viking 2's orbiter was the first to go offline. A few months after taking the first photos of the Martian moon Deimos, the machine developed a gas leak and began to lose maneuvering fuel. It was turned off in July 1978, after completing over seven hundred orbits.

The Viking 2 lander and remaining orbiter operated successfully until 1980. However, in April, the lander suffered battery failure and stopped transmitting back to Earth. Then, in August, the Viking 1 orbiter was shut down after exhausting its own supply of maneuvering fuel.

That left the Viking 1 lander as humankind's final emissary on Mars. With its nuclear fuel supply of plutonium-238, which had a half-life of 87.7 years, the ninety-day mission could, in theory, have gone on perhaps another decade or more (the thermocouples–devices that changed heat into electricity–would have degraded long before the plutonium fuel was exhausted). However, in 1982, as part of routine "housekeeping" operations, a string of erroneous computer code was sent from JPL to the lander, inadvertently instructing it to rotate its radio dish away from Earth. By the time the error was realized, it was too late; during the next communication attempt, when the lander should have had a straight-shot to talk to Earth, it did not respond. There were efforts to reflect signals off the ground nearby, but these were not successful. This was not how the mission planners wanted things to end—naturally, they preferred to shut down the spacecraft themselves, via a final command. There was something particularly bothersome about a lander lost in its prime. To this day, Viking 1 sits on the surface of Mars with its radio dish aimed at the red sand, as if awaiting its final instructions. The spectacular mission of the Vikings was over, and Mars would remain silent and unwatched for the next fifteen years.

MARTIN MARIETTA CORPORATION

6801 ROCKLEDGE DRIVE
BETHESDA, MARYLAND 20034
TELEPHONE (301) 897-6101

J. DONALD RAUTH
PRESIDENT

March 9, 1977

Dear Shareholder:

The National Aeronautics and Space Administration has awarded Martin Marietta the maximum performance fee for its accomplishment in Project Viking, the exploration of Mars.

The amount of this award, $14.8 million, is gratifying, of course, but hardly more so than that it represents 100 per cent of the sum available to us for accomplishment in one of the most difficult and most sophisticated projects ever undertaken.

I have conveyed again my deep feeling of admiration to literally thousands of dedicated men and women in the Denver Division of our Aerospace company, whose singular excellence as members of our Project Viking team have led to this further recognition for extraordinary achievement.

They, and all of us, were acutely aware in the long, sometime frustrating but always challenging years of the Viking development that this was a high-risk undertaking. In the end, as the whole world knows, there came in 1976 the exhilaration of total success as the Viking opened--for the first time in human history--a close-up window through which we are obtaining our first factual knowledge of another world.

Our men and women and their exquisite machines, the Viking landers, did everything required of them--and more. They were an integral part of a NASA, industrial, and academic combination that produced a triumph of intellect and ingenuity.

Sincerely,

J. Donald Rauth

ABOVE: This 1977 letter to the shareholders of the Martin Marietta company (now Lockheed Martin) informed them that NASA had bestowed a small bonus on the company as their way of saying, "Thanks, Viking worked!"

OPPOSITE TOP TO BOTTOM: Viking 2 view of Utopia Planitia. Some dust has already settled on the lander, just a few months after touchdown. This processed image is a good representation of the true colors on Mars.

Viking 2 looks northeast across Utopia Planitia in late afternoon light.

Viking 2 captures early morning frost on nearby ground in May, 1979.

PLUCKY PATHFINDER

After Viking 1 lost contact, Mars was silent for fifteen years. It was the longest cessation in Mars exploration since Mariner 4. The Vikings served well, but there was no immediate planned replacement. As far as Mars was concerned, the failure to find confirmed signs of life seemed to take some of the wind out of NASA's sails.

OF COURSE, NASA WAS BUSY WITH A FEW OTHER PROJECTS that occupied much of its time and funding. Voyager 1 and 2 were sent on a decades-long tour of the outer solar system and had taken center stage.

By the time Voyager 2 flew past Neptune in 1989, all the planets in the solar system save Pluto had been charted. During the intervening years, other craft had also been dispatched toward Mars. The Soviet Union had contined its quest for a successful Mars mission. The Soviets launched two probes, Fobos 1 and Fobos 2, in 1988. The mission was intended to conduct orbital observations of Mars, and to fly landers to one of its moons, Phobos. In what was becoming an uncomfortable pattern, Fobos 1 was lost during launch and Fobos 2 just prior to deploying its lander.

However, the Soviets were not alone in their frustrations. A NASA mission called Mars Observer was launched four years later in 1992 and met with failure. Communication with the orbiter was lost just three days prior to arrival at Mars, apparently due to leaking fuel throwing the craft into a spin. Subsequent investigation focused on the hardware: the orbiter had been derived from an Earth satellite and repurposed for use around Mars as part of NASA's early 1990s cost-saving measures. A repurposed "off-the-shelf" satellite was not a good choice for a deep-space mission and has not been tried in this way again since.

SUCCESS AFTER FAILURE

Other solar system exploration efforts were more successful. The Galileo mission to Jupiter was launched in 1989, after long delays due to the loss of the space shuttle Challenger, and was eventually delivered to orbit by Atlantis. While its transmitter was compromised by a partial antenna deployment failure, it nevertheless achieved its major mission objectives.

An array of spacecraft had also been sent to Venus, returning stunning results. NASA's Pioneer Venus had met with mission success in 1988, both in orbit around the planet and via dropped atmospheric entry probes. The Soviet Union scored with its Venera and Vega missions, including the return of the first pictures from the hellish planet's surface in 1981. Finally, the US Magellan orbiter succeeded in mapping most of the surface of Venus between 1990 and 1994, using radar to penetrate the dense envelope of clouds surrounding the planet.

But, as illuminated by both Soviet and US failures, exploring Mars remained a challenge. With the Soviet Union unraveling politically and economically in the late 1980s, efforts to reach Mars came

ABOVE: Mission patch for NASA's Mars Pathfinder program.

OPPOSITE: Technicians in the Spacecraft Assembly and Encapsulation Facility-2 (SAEF-2), at Jet Propulsion Laboratory prepare to close the metal "petals" of the Mars Pathfinder lander. The Sojourner small rover is visible on the forward of the three petals.

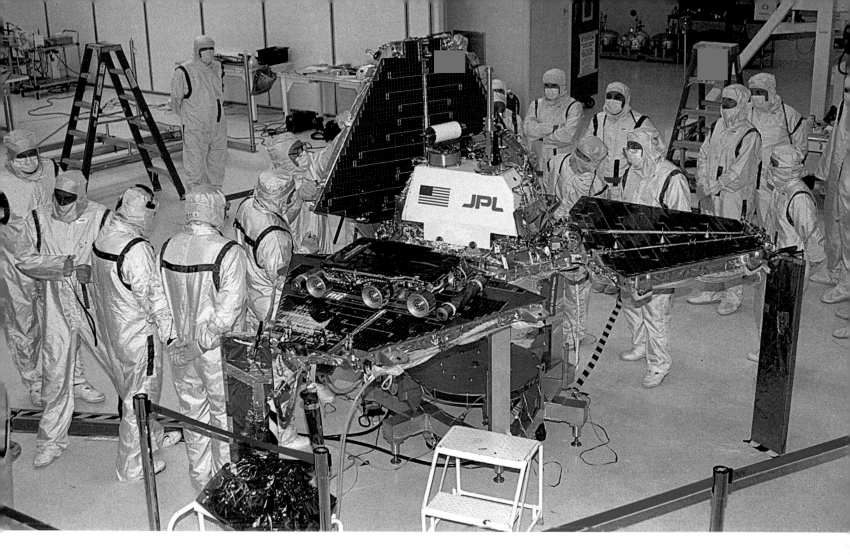

largely to a standstill for over a decade, with just one additional (failed) attempt in 1996 by the Russian Federation. In the US, an increasingly risk-averse and thrifty NASA was actively seeking less expensive ways to explore the solar system. When Daniel Goldin took the helm as NASA administrator in 1992, he initiated a new approach that he called "faster, better, cheaper." What he wanted was a larger selection of inexpensive missions that would both save costs and spread risk. Of course, ask any aerospace engineer in private if "faster, better, cheaper" made sense, and they were likely to say, "Sure, pick any two…"—but FBC was the mantra, and from this a number of smaller robotic missions were spawned, including one modestly sized Mars spacecraft called Pathfinder.

In stark contrast to the expensive and ambitious Viking program, Pathfinder was lightly funded and took only a few years to prepare. It was a single, unorthodox spacecraft with modest goals, but once JPL got the green light it was full speed ahead for the new mission.

PATHFINDER LEADS THE WAY

Pathfinder, with a three-year development schedule and a budget of just $150 million, was part of NASA's new Discovery Program, which was intended to spawn smaller, less expensive robotic missions with limited goals. After the billion that had been spent

on Viking, which in 1997 dollars could easily have been more than double that in adjusted costs, Pathfinder seemed like it would be an absolute bargain... if it worked.

JPL had long used outside companies to help build and run its missions under contract. But this takes time, something the Pathfinder mission had very little of. In an unusual turn, the mission was built and operated entirely in-house at JPL. It consisted of a small lander, weighing in at just 500lb (226kg), and a tiny toaster-oven sized rover that tipped the scales at under 25lb (11kg). When you realize that one un-fueled Viking lander weighed about 1200lb (544kg), Pathfinder seems all the more remarkable.

The team behind the Pathfinder mission was uncharacteristically young and lean. They worked fast, tested only what absolutely needed testing, and eschewed paperwork. With a three-year timeline (Viking's had been over a decade long), there was little time for recordkeeping and paper trails. As one lead engineer put it, "There was no time for memos… we flew largely under the radar." *Esprit de corps* was incredibly high.

The mission design was clever. Due to the tightly constrained budget, the Pathfinder planners were not allowed the luxury of entering Martian orbit and choosing a landing site at their leisure. The rocket, a far smaller (and cheaper) Delta II, as opposed to

Viking's massive Titan III, only had enough power to send the tiny payload in a direct trajectory to Mars—it could not carry the fuel and engines needed to enter orbit around the planet. This was almost like dispatching the spacecraft to Mars in a giant cannon, with the bullseye painted on the Martian surface—but given the budgetary and propulsive constraints, was the only viable option. This became the trajectory of choice for all future Mars landers.

The Pathfinder team utilized the same Mars maps that the Viking planners had used to plan their own landings two decades earlier. Nothing new had orbited the planet since the Viking orbiters fell silent, so planners returned to the Viking and Mariner 9 maps for landing site selection. Once again, a lot of intuition and guesswork were applied. One advantage they had was that the mission scientists had been able to compare the Viking orbiter pictures to the images returned from the surface in the 1970s, and this helped to understand a bit more about what they were looking at in the orbital maps, giving them ground-level context. However, making an intelligent choice of a safe but geologically interesting landing site remained a long shot. There was still much unknown territory on the surface of that planet, and coming down safely was going to be tough—especially for such a tiny machine.

The young engineers studied the Viking designs long and hard, then promptly put them aside. Even with a decade of engineering improvements, including vast gains in computer capability, the lander would still, in essence, be landing blind, just as the Vikings had. And given the vastly smaller budget, they knew they needed a completely new approach. How could they take some of the risk out of the equation?

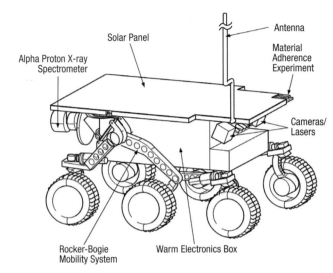

OPPOSITE: The rover, Sojourner, was the first machine on Mars after the 1976 Viking landers. In contrast to their 1260lb (572kg) weight, its weight was just over 25lb (11.5kg). Nonetheless, it was quite a capable rover.

RIGHT: The airbag landing system was revolutionary, yet gave the engineers unending troubles during testing. In use it worked so well that it was utilized for the next rovers, MER, as well.

BELOW: The landing sequence for Pathfinder: after it entered the atmosphere, a parachute slowed the lander until radar told it that the ground was close enough to lower the lander on a rope. Rockets then fired, the airbags were inflated, and the rope cut. Pathfinder bounced to a landing.

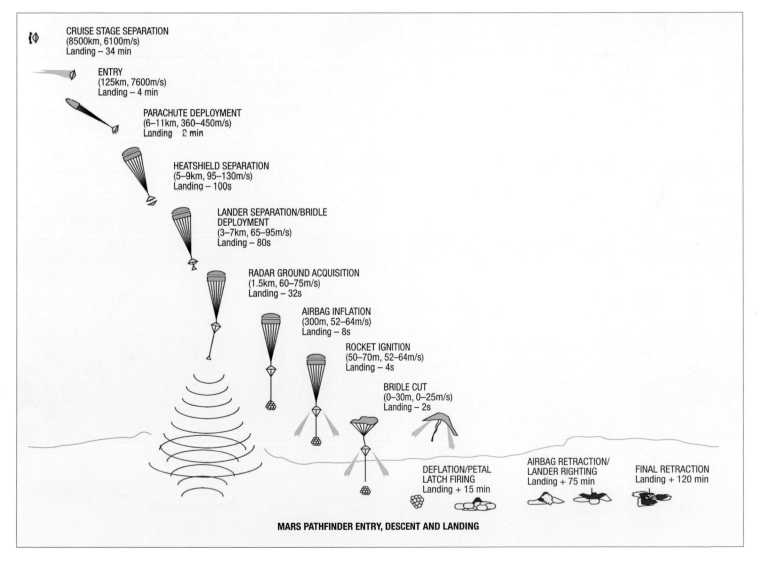

CRUISE STAGE SEPARATION
(8500km, 6100m/s)
Landing – 34 min

ENTRY
(125km, 7600m/s)
Landing – 4 min

PARACHUTE DEPLOYMENT
(6–11km, 360–450m/s)
Landing – 2 min

HEATSHIELD SEPARATION
(5–9km, 95–130m/s)
Landing – 100s

LANDER SEPARATION/BRIDLE
DEPLOYMENT
(3–7km, 65–95m/s)
Landing – 80s

RADAR GROUND ACQUISITION
(1.5km, 60–75m/s)
Landing – 32s

AIRBAG INFLATION
(300m, 52–64m/s)
Landing – 8s

ROCKET IGNITION
(50–70m, 52–64m/s)
Landing – 4s

BRIDLE CUT
(0–30m, 0–25m/s)
Landing – 2s

DEFLATION/PETAL
LATCH FIRING
Landing + 15 min

AIRBAG RETRACTION/
LANDER RIGHTING
Landing + 75 min

FINAL RETRACTION
Landing + 120 min

MARS PATHFINDER ENTRY, DESCENT AND LANDING

THE LANDING CONUNDRUM

After much wrangling and back-of-the-napkin brainstorming, one design worked its way to the top of the stack. They would, in effect, design the lander in such a way that it would find a safe spot to settle *after* it had reached the surface. The lander would be enveloped in an assembly of inflatable, protective airbags, and once released from the parachute would bounce, and then roll, until it settled in a clearing. Once at rest, the airbags would deflate and the lander's protective housing would open, automatically righting itself—that part of the design was somewhat reminiscent of the Soviet Mars 2 and Mars 3 landers from 1971. Only then would the camera deploy and the rover depart its landing stage.

This was the plan. It was new, daring, and innovative as hell. But this new approach to landing on Mars did not sit well with many of the older engineers at NASA headquarters.

In a now infamous meeting between Pathfinder's senior engineers and managers and the "old heads" at NASA, many of the seasoned veterans of such projects as the Surveyor lunar landers, the Apollo program, and even Viking, were not happy. When the Pathfinder design was unveiled there was a stunned silence in the room. Rob Manning, JPL's chief engineer on Pathfinder, recalls Cadwell Johnson, one of the principal designers of the Apollo spacecraft, reacting to the presentation: "Listen, buster, don't tell me how to land on another planet! This is a stupid idea that's never gonna get off the ground!" He still remembers the day vividly. "The Viking people were rolling their eyes and saying 'You're gonna do *what*?'" he says.

"At that point we had told them how high we would be bouncing—probably 50 to 75ft [15–23m] above the surface of Mars," Manning adds. "We were testing it to be able to bounce 100ft [30m]. They just thought we were nuts."

The headquarters review did not go well. But after a lot of wrangling, the Pathfinder team eventually got the sign-off and was allowed to proceed. Now everything had to be tested, and cheaply.

Airbags were sewn together and inflated with weights inside to simulate the mass of the spacecraft, then test-dropped on a variety of surfaces. They ripped and tore, again and again. Sizes, pressures, and configurations were experimented with until the designers got it right, but there was not much margin for error. The engineers simply did not have the money, or time, to test the airbags as much as they would have liked.

The parachute provided similar challenges. Only twice before had American parachutes been used on any other planet, to deliver the Viking landers to Mars. With much less money to spend, and a necessarily faster entry speed into the Martian atmosphere—about fifty per cent faster than Viking's—the Pathfinder team tested as many design variations as they could afford. The chosen design

eventually stopped ripping and snarling, but again, the margin for error was not comforting to an engineer's orderly mind.

And then there was the tiny rover. Named "Sojourner" (after Sojourner Truth, an American campaigner for slaves' rights in the 18th century), the rover was simple in design but, as the first of its kind, was facing vast unknowns. It too needed testing. With an unrelenting eye toward the tight budget, a lumber enclosure was built in an empty room at JPL and filled with sand from a local playground supply store. The rover was tested in the same sand used at the neighborhood elementary school. It would have to do.

PREPARING FOR TAKE-OFF

Work at JPL continued, with the already tight schedule demanding more and more time and resources right up until the spacecraft was due to depart for Cape Canaveral in Florida. As the first mission to land on Mars since Viking, and the first wheeled robotic

BELOW: The path between Earth and Mars is much farther than one might think. At their closest the planets can be within 35 million miles (56 million km) of each other, but spacecraft travel a curved trajectory to reach the planet, so the distance covered is well over 300 million miles (483 million km).

OPPOSITE TOP: The Sojourner rover is fitted to the pedal on Mars Pathfinder.

OPPOSITE BOTTOM: The family resemblance: Mars Pathfinder's Sojourner rover to left, the MER rover Spirit to right. Curiosity would tower over both.

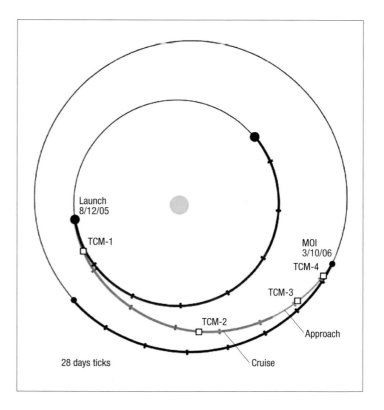

Launch
8/12/05

TCM-1

MOI
3/10/06

TCM-4

TCM-3

TCM-2

Approach

28 days ticks

Cruise

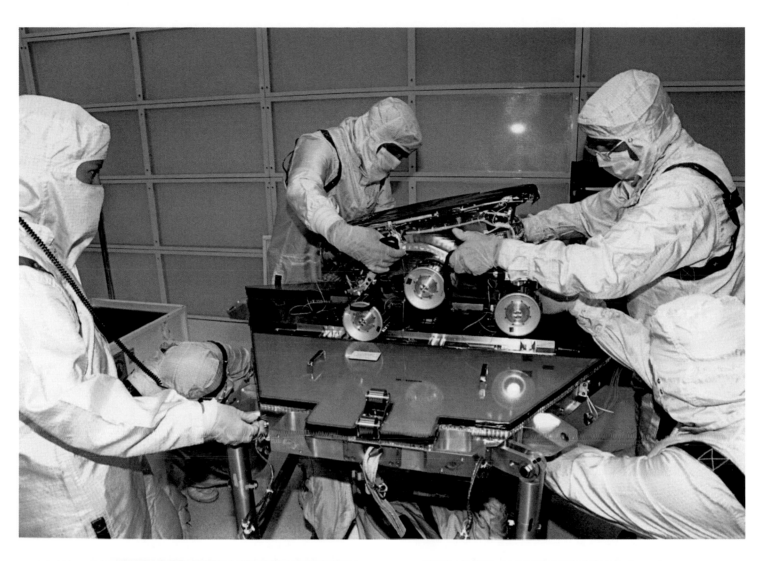

MARS PATHFINDER

MISSION TYPE: Mars lander, Mars rover
LAUNCH DATE: December 4, 1996
LAUNCH VEHICLE: Delta II
ARRIVAL DATE: July 4, 1997
MISSION TERMINATION: September 27, 1997
MISSION DURATION: 2 months, 23 days on Mars
SPACECRAFT MASS: 1020lb (463kg)

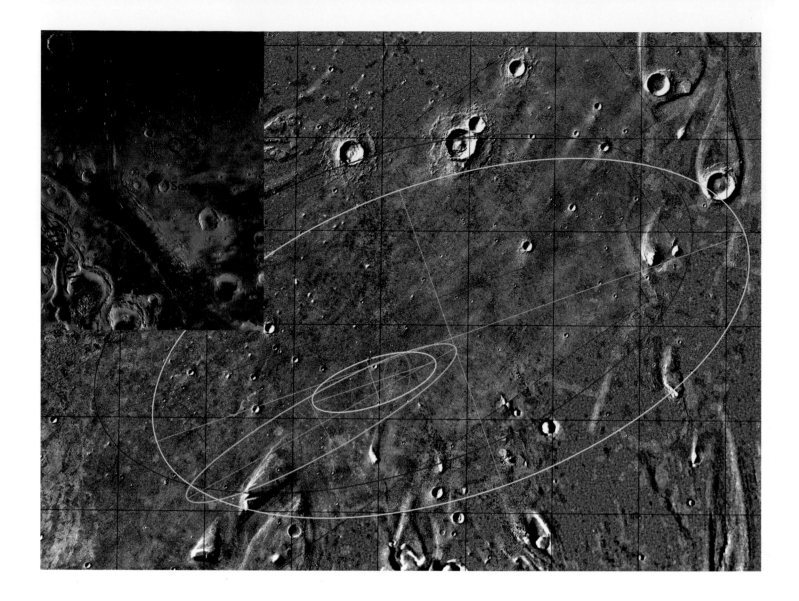

machine to land anywhere off-earth since the Soviet Lunakhod of the 1970s, the stakes were high. And though there had been every intention of outsourcing much of the fabrication to a NASA contractor, there simply had not been time to do so—both the lander and rover were actually built at JPL by NASA engineers and fabricators. It was the ultimate in-house undertaking, with the aero-shell, airbags, and landing rockets the only major components outsourced.

As Pathfinder prepared to launch in December 1996, some at NASA headquarters began to worry. Pathfinder had not received a lot of scrutiny since the contentious design review, but now that it was scheduled to fly soon, some wondered if all the "i's" had been dotted and the "t's" crossed. However, with the fast-and-

furious pace of the preparations, there was not the usual tonnage of paperwork to be examined. Meetings were held, JPL engineering teams consulted, and expectations adjusted. The mission was meant to be primarily an engineering proof-of-concept anyway—if it worked at all that would be a success, and any science performed beyond the primary mission was optional. The lander was expected to last a month, the rover a week. Anything else was a bonus.

This did not, however, lessen the stress on the engineers and mission managers at JPL. NASA headquarters had their set of expectations, and the Pathfinder team at the laboratory had their own tough standards to meet. The next few months would demonstrate the measure of their Mars capabilities, and chart the course for future landings.

ABOVE: Pathfinder pioneered the airbag landing system. This landing ellipse in Ares Vallis was not far from viking 1's landing site in Chryse Planitia. The final landing ellipse (in pink) was 124 miles (200km) by 43 miles (70km).

OPPOSITE: The moment of truth – after a high-speed atmospheric entry, the Pathfinder lander, enclosed in its protective Vectran airbags, impacts the surface. Artist's impression.

Rob Manning

Pathfinder Chief
Engineer

Rob Manning joined JPL after graduating Caltech in 1981 with a degree in electrical engineering. After a decade of working on the Galileo Jupiter probe and Cassini's mission to Saturn, Manning was recruited to work on a new, smaller mission. At first, he had his doubts about Pathfinder: "When I first saw the design, I thought, 'JPL has no skill in this, it's been so many years since we've actually landed on a planet, we hadn't really done a lander at JPL since Viking in 1976!'"

But after a recruiting call from Tony Spear, the project manager, Manning changed his mind. "He was a mechanical engineer, and I was an electronics and software systems engineer, and we just hit it off right away, so he hired me as the chief engineer on the project." Soon Manning had taken charge of

the most vexing part of the entire mission— EDL, or entry, descent and landing. It was not a job for the faint of heart.

"Tony Spear, the project manager, wanted to outsource EDL," Manning recalls, "but we didn't have much time before we launched, and it was very complicated."

Manning wasn't even sure of a way to explain to an outside company how to bid on the job, much less how to help them design it. "I couldn't imagine writing a spec to tell people how they had interface with an airbag, for example, so we talked Tony into letting us continue the process as we had been—to do it here. And that worked out great, and could not have been achieved without incredible teamwork."

PATHFINDER TRIUMPHANT

Pathfinder thundered out of Cape Canaveral on December 3, 1996. There was no backup—this single spacecraft had to work, or there was no mission. But everything went off without a hitch, and Pathfinder headed directly for Mars. There would be no dalliance in Mars orbit, no surveying the ground for those second-guessing the landing site. Those decisions had been made sometime ago, and Pathfinder's engineers knew exactly where it was headed, for better or worse.

ON JULY 4, 1997, PATHFINDER landed in a region named Ares Vallis, just 520 miles (836km) from Viking 1's landing site, in Chryse Planitia. Ares Vallis was thought to be a region affected by ancient water flows, and thus an area of great interest to the geologists. It had been exhaustively studied from orbital photographs, and offered a reasonable expectation for scientific return tempered by a strong desire for the safe delivery of the spacecraft.

Pathfinder made its high-speed entry into the Martian atmosphere, flying autonomously due to the long radio delay. The small onboard computer timed its course changes perfectly and adjusted the trajectory of the tiny spacecraft as the heat shield protected it from the blistering temperatures of entry. The parachute was deployed at supersonic speeds of just over 800mph (1287km/h), and did not tear upon inflation as it had in so many tests. This slowed Pathfinder to a more leisurely 160mph (257km/h).

About 3 miles (5km) from the surface, Pathfinder was winched rapidly down from the slowly descending rocket pack. Dangling

from 65ft (20m) of line, about 1000ft (305m) above Mars, the airbags were instantaneously inflated by hot gas from a set of small rocket motors. With just a few hundred feet remaining, braking rockets were fired to aggressively slow the craft, then, at less than 70ft (21m), Pathfinder was dropped free and fell to the surface.

The first bounce from the 40mph (64km/h) impact was about 45ft (14m) high, with each succeeding rebound a bit less each time. There were fifteen bounces recorded by the onboard sensors; there may have been more. Soon Pathfinder rolled to a stop, and the airbags were deflated. The lander sent one critical message home: it had survived its unorthodox descent and was down, and safe, on Mars.

PATH FOUND

As their forebears had twenty-one years previously, the members of the landing team erupted in jubilation. And the world was watching—this time from millions of locations across the globe. As

the first planetary landing in the age of the internet, anyone was able to tune-in, and many millions did. It was the largest demand the JPL servers had ever experienced, and they were strained to the limits by the load. The iconic image of an ebullient, bearded mission leader Rob Manning, one fist held high in triumph, was soon spread across the web in an early manifestation of viral imagery.

JPL controllers checked out the health of the spacecraft, which was excellent. The landing had occurred at about 3am local Mars time. The deflated landing bags were retracted, and the protective sides of the enclosing structure opened like the petals of a flower. After sunrise, the first images were sent back to earth and weather measurements were taken. It must have felt oddly familiar to those in attendance who had worked on the Viking mission.

On the second Martian day, called a sol, the small rover was unclamped from the landing stage and prepared to roll down onto the Martian surface. This was another risky phase—the rover had to navigate a steep ramp to get to the sand below. There had been much concern about all the things that could go wrong during this seemingly simple maneuver, and tensions were high. Becoming hung-up on the ramp, or tangled in the now-deflated airbags that surrounded the rover, would be a major setback.

The rover traveled down the ramp incredibly slowly—caution was the watchword. Sojourner had an optimum speed on flat ground of about 0.5 inch (13mm) per second, but the ramp was traversed much more slowly. It took the better part of a day to get all six wheels on to Martian soil. At dusk, the rover deployed its Alpha X-ray Spectrometer and sniffed the soil right where it was parked at the base of the ramp. The results were about as expected—the soil generally matched the data from the Viking missions two decades earlier.

On Sol 3, Sojourner set out for its first drive. The rover headed to a nearby rock target, which the team had named, somewhat whimsically, Barnacle Bill. The team of scientists on the mission were allowed to come up with their own names for the objects they encountered, so everything from cartoon characters to original silly concoctions were invoked. Barnacle Bill was a scant 15in (38cm) from the ramp—not much of a distance to travel, but far enough for a first outing. Examining the rock with the APXS spectrometer on the rover's small arm took ten hours, and revealed it to be volcanic in origin—an andesite, or a rock that had most likely been volcanically formed, re-melted, and re-formed over time.

OPPOSITE LEFT: The Sojourner rover is seen in its flight configuration, the day after landing. Once the restraints were released, it would "stand-up" on its suspension and roll down a ramp to the surface. The deflated airbags are seen surrounding the rover.

OPPOSITE RIGHT: Pathfinder's rover, Sojourner, was about the size of a microwave oven and weighed just 25 pounds. It was the first wheeled machine on Mars.

ABOVE: On sols 8–10 of the mission, Pathfinder's camera imaged this panorama of the landing site. The Sojourner rover is inspecting Barnacle Bill, the boulder to the right.

Meanwhile, the lander took its first "monster pan," a 360-degree photographic montage of the landing zone. It was the first of many inspections of the surrounding terrain, each of which revealed new features. The landing zone was as the scientists had suspected—a scene of ancient, catastrophic flooding, which had taken place untold eons ago. Water-transported deposits were visible not far off, and the nearby rocks were clearly of different compositions and, most likely, origins. It was, as one scientist called it, a "geological grab-bag." The team was giddy with excitement for weeks.

The second rock the rover visited—dubbed Yogi—was examined with the APXS and found to be of different origins than Barnacle Bill. Yogi was a basalt—an older and more common form of volcanic rock. Close examination hinted that Yogi had been transported to its current location by water—another first for Mars exploration. This is just the kind of diversity that had been hoped for, and to find it on the first planetary mission with a functional rover was a sensation.

A CHALLENGING MISSION

The weeks went by with new discoveries coming with stunning regularity. Sojourner was functioning well beyond the seven-day primary mission, and clearly the Pathfinder lander would outlast its thirty-day warranty. Yet the mission was not without its challenges.

In mid-August, the computer on the lander decided to reboot without warning. The team at JPL watched with mounting concern as it restarted spontaneously—it simply shut-down and rebooted. The computer was an old design even at the time of flight, a 2.5 Mhz IBM RAD6000 (a militarized version of the old Macintosh PowerPC 750 chip), with 128 megabytes of RAM. There was great relief when it came back online hours later… but another problem soon became apparent.

While the lander had been out of contact with the ground, the rover had gotten in trouble. Able to drive short distances autonomously, one side of the rover had driven onto a wedge-shaped rock (they named it, appropriately, Wedge), and the resulting tilt exceeded preset safety parameters. Sojourner stopped short and awaited orders from Earth. After a hurried conference and some on-the-ground testing, the engineers decided on a course of action and radioed instructions up to the rover. Sojourner slowly backed off the rock, reoriented itself and headed to another area they called Rock Garden. A week later, it arrived in the jumbled collection of geological specimens.

LEFT: An artist's impression of the Pathfinder spacecraft on the surface of Mars, with the Sojourner rover heading off towards a target rock. Note that even the area where the rover had been secured for flight, between the two ramps, is covered with a solar power panel. Efficiency was key to the successful mission.

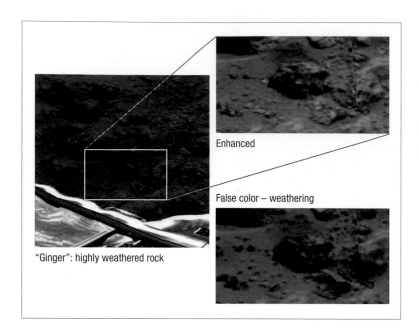

Enhanced

False color – weathering

"Ginger": highly weathered rock

Sojourner spent the rest of the mission exploring rocks and soil features within a 30-ft (10-m) radius of the Pathfinder lander. As the geological and chemical evidence piled up, a more detailed picture of Mars was quickly evolving. True to the indicators from orbit, readings showed that the planet had clearly experienced water washing over its surface in the past. They were driving across part of a flood plain. But where had all that water gone?

The water question would have to wait for another mission. The computer on Pathfinder continued to experience increasingly frequent and concerning glitches, and the onboard battery was losing its storage capacity. Martian winter was upon them, and with it came increasingly cold temperatures. With reduced storage capacity, the heaters in the lander were not operating properly, and the electronics were getting too cold in the long Martian nights.

THE END OF THE QUEST

Mars is a harsh environment and tough on spacecraft, and Pathfinder's days were numbered. On September 27, 1997, the final transmission came down from the lander and then it went silent. Repeated requests from Earth went unanswered, and it seemed that the mission was over. There were more attempts to restore communication until the end of November, but without

ABOVE LEFT: A Pathfinder image of a rock called Ginger is to the left. The images to the right are color-enhanced to display additional detail.

LEFT: Image of the rock Chimp taken on September 15 by the Sojourner rover's front cameras. Intricate texture can be seen on the rock. Trails produced by wind, oriented from lower right to upper left, are seen next to small pebbles in the foreground.

success. The end of the mission was officially announced on November 5, 1997.

In Pathfinder's eighty-five days operating on the surface of Mars, it performed fifteen separate chemical analyses of rocks and soil, returned 550 images from the rover and 16,500 images from the lander's camera mast. Sojourner lasted eleven times' its expected seven-day lifespan, and the lander triple its thirty-day warranty.

Perhaps the crowning accomplishment of the mission was to provide ground-truth verification, and refinement, of what had been suspected when looking down from orbit: unquestionably, in its past Mars had been warmer and far, far wetter than it was today. The evidence was everywhere. The story long held in the rocks and soil had been cracked open, and "follow the water" would become NASA's motto for all Mars exploration henceforth. Not bad for a low budget program that barely made its deadline to the launchpad.

Even as the now-lifeless Pathfinder began to accumulate a coating of red dust with the encroaching Martian winter, JPL's engineers were busy at work on their next wheeled Martians— the Mars Exploration Rovers. However, long before those far more sophisticated machines would arrive at the Red Planet, another robotic visitor would arrive to look down from above with unprecedented clarity: Mars Global Surveyor.

ABOVE: A composite of Pathfinder images form July, 1997 shows a Martian sunset. The sky color is accurate; the terrain is has been lightened to bring out detail.

RIGHT: A schematic view of Pathfinder's landing site, created by processing down-looking images taken from the lander's mast cameras. The rocks investigated by the rover are indicated via red boxes.

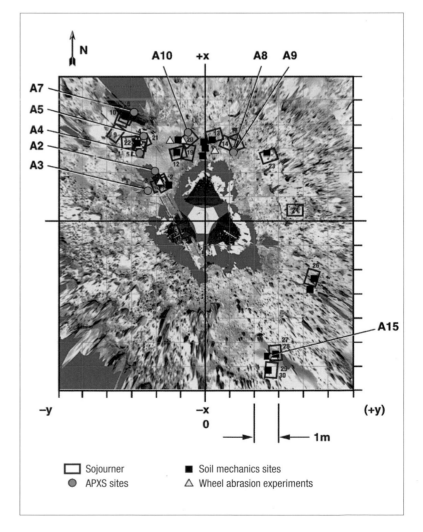

FROM ON HIGH:
MARS GLOBAL SURVEYOR

Though launched slightly before the Pathfinder mission in November 1996, the Mars Global Surveyor (MGS) orbiter actually arrived at the planet later, on September 11, 1997. Like the Vikings and Mariner 9 before it, MGS fired braking thrusters to fall into orbit around the red planet. The orbiter settled into an orbit that was 163 miles (262km) from the planet at its low point, but over 33,000 miles (53,108km) at its high point—an unusual orbital trajectory. Why the oddball ellipse?

THE STORY BEGINS OVER TEN YEARS previously, about the same time the last Viking lander lost contact. In 1984, NASA was looking for a new Mars orbiter to pick up where the Viking had left off—another orbiter with newer and improved instrumentation was the logical next step. Viking, with two orbiters and landers, had been expensive and another mission of that magnitude was out of the question. A single, reasonably priced orbiter seemed feasible. Another bonus was the fact that the space shuttle was delivering payloads to orbit, and NASA might be able to launch the new spacecraft from its cargo bay.

So far, so good. But as always, other considerations began to emerge, and the program, which would ultimately be called Mars Observer (MO), began to become distinctly bogged down. In order to save money, the spacecraft was based on a design for an Earth-orbiting satellite that would be repurposed for a journey to Mars. Certain systems would have to be modified for the deep space voyage, but the basic design would be the same, which would, in theory, save money.

ECONOMY SPACE TRAVEL

The decision to launch the craft from the space shuttle had come down from NASA headquarters. To optimize the hoped-for savings associated with regular flights of the shuttle, just about everything space-bound was supposed to be carried aloft in the orbiter's cargo bay. However, when Challenger exploded seventy-three seconds after liftoff in 1984, that order was rescinded. In the future, satellites and planetary missions—as well as most military

ABOVE: The winning logo in the competition to design the Mars Surveyor 98 project logo.

OPPOSITE: This image is a processed composite of numerous images from Mars Global Surveyor taken in 2006. The Tharsis region, home to Mars' largest volcanoes, is in the center.

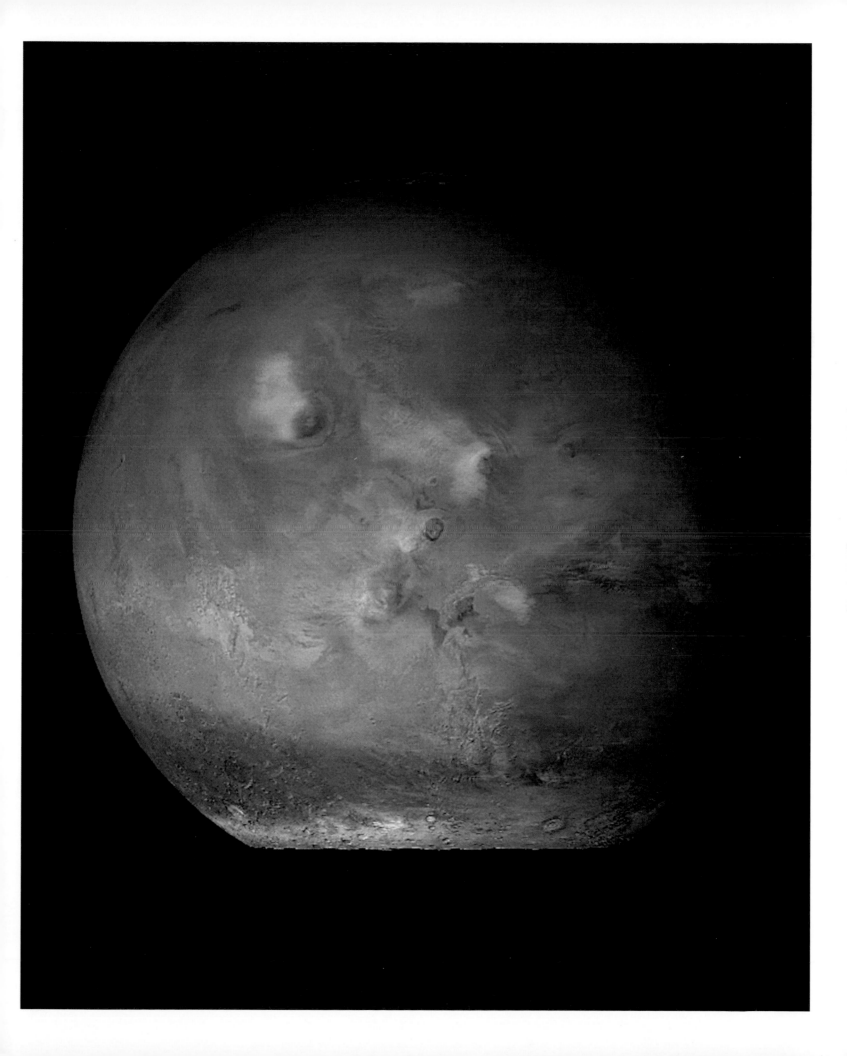

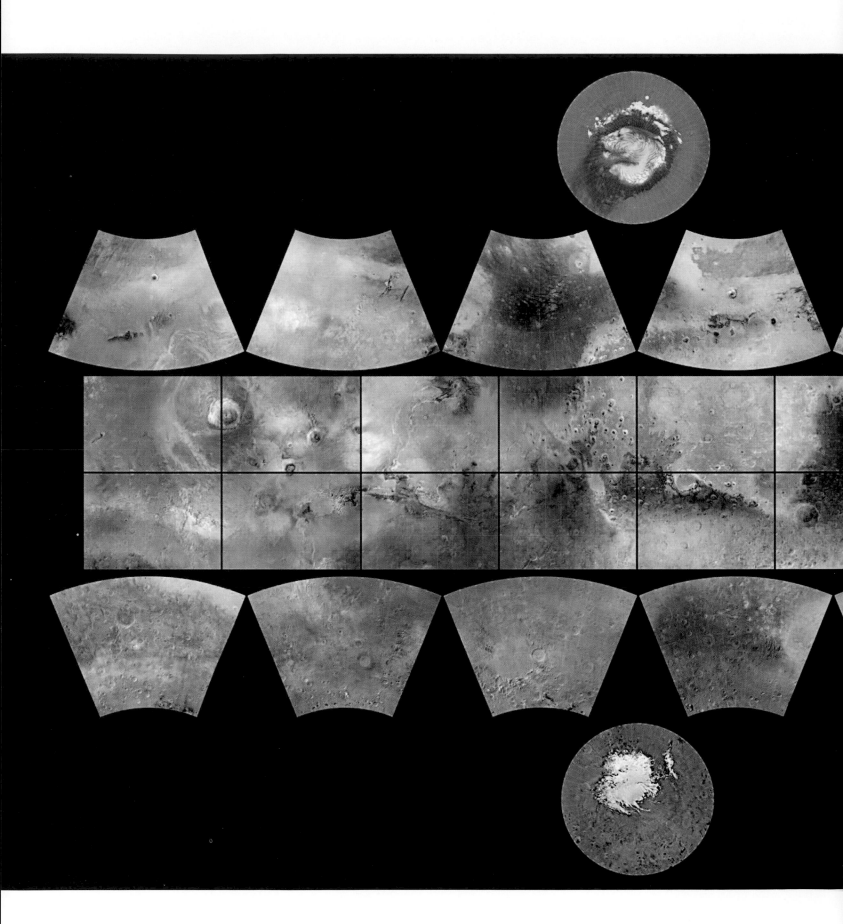

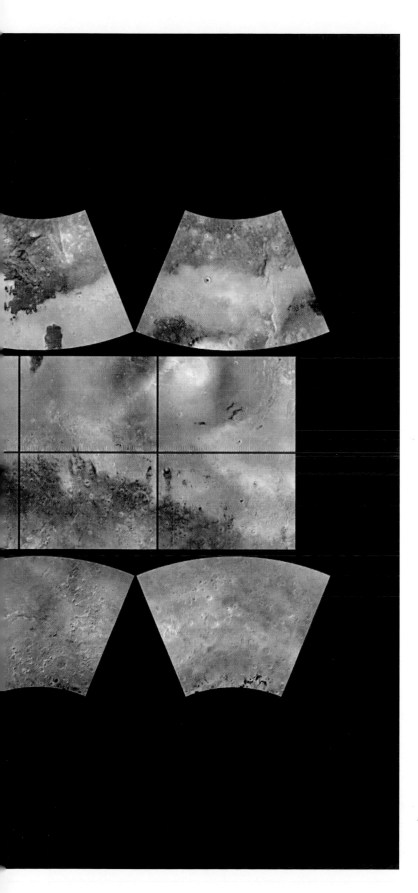

missions—would fly on expendable rockets, as they had for the preceding twenty-five years.

One problem with this approach was that one-use rockets are very expensive to operate, and Mars Observer was no lightweight. At 2244lb (1020kg) it would have to fly on a costly Titan III, just as the Vikings had done. Yet more corners were cut in the program to accommodate the rising costs.

In September 1992, Mars Observer finally left the Cape in a cacophony of smoke and thunder. Things generally went well for almost a year, as the spacecraft closed in on Martian orbit. Then, in August 1993, shortly before the burn that would place Mars Observer into that orbit, communication suddenly ceased. The spacecraft fell silent and stayed that way.

After an excruciating examination of the evidence and a detailed look at the project's development history, it was determined that a leak had developed in the fuel system, resulting in a catastrophic, mission-ending explosion. It was thought that the repurposing of an Earth-orbital design and hardware was in part responsible for the loss of the spacecraft. There are cost savings, and then there are false economies. This mission had been predicated upon the latter.

Almost immediately NASA laid out a new initiative—the Mars Exploration Program (MEP). The MEP combined newly specified goals for Mars exploration with a strong need to recover—and quickly—from the embarrassing Mars Observer failure. The results were the inexpensive Pathfinder lander/rover, and an entirely new design for an orbiter, the Mars Global Surveyor (MGS).

MGS would attempt to salvage much of what the MO program was intended to do, but at less cost—about $154 million versus over $800 million—and with about half the mass. Some flight hardware was pulled off the shelf from Mars Observer spares, but the overall design was new—and intended from the beginning to be a Mars orbiter.

Due to MGS's smaller mass, the less expensive Delta II rocket was used to launch the mission. Like the Titan, it was a former ICBM nuclear missile launcher, repurposed for less lethal purposes. However, unlike the Titan, which developed well over 1.3 million pounds of thrust at launch, the basic Delta II generated just under 800,000 pounds of thrust. MGS would not enjoy the luxury of carrying a large enough rocket motor to brake into Martian orbit. The engineers and mission planners would need to come up with some new and ingenious way in order to slow the spacecraft and keep it from zinging past Mars like one of the early Mariner flybys.

LEFT: This image is an update of a Mars map created with Mariner 9 images in 1979, using new imagery from Mars Global Surveyor. All aspects of Mars can be seen in this detailed map.

MARS GLOBAL SURVEYOR

MISSION TYPE: Mars orbiter
LAUNCH DATE: November 7, 1996
LAUNCH VEHICLE: Delta II

ARRIVAL DATE: September 12, 1997
MISSION TERMINATION: November 2, 2006

MISSION DURATION: 9 years, 1 month, 21 days at Mars
SPACECRAFT MASS: 2272lb (1030kg)

HITTING THE BRAKES

There had been talk for years about a method of slowing a spacecraft after its high-speed interplanetary cruise called "aerobraking." In brief, the spacecraft's trajectory would be adjusted to dip slightly into the atmosphere of the target planet, and the resulting aerodynamic drag would, in theory, slow it down enough over successive passes to pull it into proper orbit. There were potential downsides of course (there are tradeoffs in spaceflight)—any errors in aiming, or a misunderstanding of the density of the target atmosphere, could cause the craft to overshoot or burn up. Also, it was not entirely clear how robust a spacecraft would have to be in order to survive such a maneuver. The technique had been tried only once before.

When the NASA Magellan mission to Venus was nearing the completion of its orbital research in 1993, the agency decided to try an aerobraking experiment. After all, most of the important work was done, and the craft was entering its fifth mission extension. An added incentive was that Venus has a nice, thick atmosphere that should provide clear results indicating the success or failure of slowing the spacecraft.

The trajectory of the spacecraft was altered, Magellan dipped into the Venusian atmospheric blanket, and over the course of two months, the spacecraft's orbit was altered from an elliptical one to a generally circular one, just as planned.

In the case of MGS, aerobraking was incorporated from the beginning. The braking rockets would be far smaller than those used on the MO probe, but would be able to slow MGS sufficiently to drop it into a hugely elliptical one, 163 × 33,000 miles (262 × 53,108km) long, which required a lot less braking power. Aerobraking, accomplished with the very, very thin Martian atmosphere, would have to do the rest. It was both an elegant solution, but a big gamble. NASA's choices were, however, limited—it could wait until a larger spacecraft could be built and a larger rocket bought—or go ahead with this once-tested technique. So aerobraking was the chosen solution.

The MGS spacecraft was the first to be designed with this approach specifically in mind. With its 40ft/12m-wide solar panels, MGS would create enough drag in the Martian atmosphere to alter its orbit substantially within six months of its arrival. At least that was the plan.

MGS IS GO!

MGS launched on November 7, 1996, almost a full month before Mars Pathfinder. It arrived at Mars in September 1997. During the cruise to Mars, however, one major technical problem had been discovered: one of the solar panels had failed to deploy, or unfold, properly. These panels were critical not just for power, but also for the correct aerobraking of the spacecraft. The solar panels were

---

MARS OBSERVER

MISSION TYPE: Mars orbiter
LAUNCH DATE: September 26, 1992
LAUNCH VEHICLE: Delta II
ARRIVAL DATE: N/A
MISSION TERMINATION: August 21, 1993
MISSION DURATION: 331 days—mission failure
SPACECRAFT MASS: 2244lb (1017kg)

---

the "wings" that would create sufficient drag to slow the craft and round out the highly elliptical orbit. One had deployed fully; the other was only about eighty percent extended. There appeared to be a problem with the bracket that supported it—it had not locked the panel into place.

But flight controllers had few options; Mars was upon them. The braking engine was fired and MGS went into its long, looping orbit of Mars, each of which lasted forty-five hours. This mission was the first to enter a polar orbit around Mars, circling over the north and south poles, a vantage point from which it could most effectively map the entire planet. Repeated maneuvers lowered its closest point to Mars, or "periapsis," to about 68 miles (109km), low enough to dip into the atmosphere and create sufficient drag to slow the craft down. The angle of the damaged solar panel was adjusted to minimize any bending or other damage that might occur due to its weakened status, and the aerobraking "dip" into the atmosphere was changed to be shallower and less aggressive. This would lessen the aerodynamic drag and slow the orbital change, but would also spare the spacecraft any more damage.

It took over a year to reach a circularized orbit, but eventually MGS was circling Mars every two hours at an altitude of about 280 miles (450km). The science team had not been idle; plenty of good data had been acquired during that time. But now the most ambitious phase of the mission could begin.

Instrumentation on MGS had been limited by size and weight requirements, and the spacecraft carried fewer instruments than had been jammed aboard the now lost Mars Observer. However, it was still a powerful suite of investigative hardware:

The Thermal Emission Spectrometer (TES) was an infrared spectrometer, particularly sensitive to minerals of interest to the researchers. A mapping altimeter, The Mars Orbital Laser Altimeter (MOLA), would use infrared laser beams to map exact distances from the orbiter to the Martian surface in precise detail.

Additional instruments included a magnetometer and electron reflectometer that would measure solar wind and help to define Mars' magnetic field (which turned out to be minimal), and the Ultrastable Oscillator for Doppler Measurements (USORS). The USORS was a very precise clock that linked in to the radio on the spacecraft, with which scientists could track minute changes in the radio signal as the probe orbited Mars, allowing for extremely accurate measurements of the planet's gravitational field.

Perhaps the most remarkable device on MGS was its camera. It was certainly the crowd pleaser on the mission. Dubbed the Mars Orbital Camera (MOC), this precision imager was built for NASA by Malin Space Science Systems in San Diego, California. Mike Malin was a former NASA employee who left the agency to follow his own path, in order to build what he considered to be a vastly superior camera to that which NASA had been using to date. He then turned around and sold his improved hardware back to NASA for use on many Mars orbital missions from MGS onward, and also retained a contract to archive and interpret the data collected by the camera. This one device went a long way toward making MGS the most capable Mars orbiter to date, providing incredible detail at a resolution of about 18in (45cm) per pixel—better than anything that had previously orbited the planet.

In its second year of operations at Mars, MGS got down to serious mapping and surface investigation from its 248-mile/399km-high final orbit with a duration of just under two hours. This period and altitude had the added advantage of allowing the orbiter to see the ground below at about the same time each day, making photographic comparisons and interpretations far more valuable.

SUPERIOR IMAGES AND KEY DISCOVERIES

Along with a wealth of data from its various instruments, the images returned were spectacular—significantly better than anything previously attained. These were studied for over a decade and resulted in a number of discoveries, including the sedimentary

nature of some of Mars' surface features. One scientist at Mike Malin's company, Ken Edgett, studied an estimated 240,000 images over a number of years, representing the entire image return of the mission, before realizing that he was indeed looking at sedimentary deposits, some up to 6 miles (10km) high. This finding—characterized as a "barn burner," and perhaps the most important of the mission by one geologist—changed the way that scientists looked at Mars and was pivotal in the design of subsequent missions.

Other discoveries during this period included confirmation of the persistent role of water in the formation of ancient Mars river deltas and even possible indications of recent water erosion in gullies etched into canyon walls. MGS was also able to take high-quality pictures of Mars' two moons, Phobos and Deimos, increasing the understanding of these tiny, secretive worlds. Martian weather patterns and atmospheric phenomena, including small cyclones called dust devils, were recorded and studied. The MOLA laser altimeter instrument provided the first true understanding of the dynamic surface topography of the planet.

MGS was also the mission that placed the search for water front-and-center into the Mars program, capturing high-resolution images of features that were clearly forged by large amounts of flowing water sometime in Mars' wet past. The mission soldiered on for nearly a decade, enjoying four extensions of its planned one Martian year (two Earth years) duration. Then, in 2006, the orbiter fell out of communication with Earth. A faint signal was received three days later, indicating that the spacecraft had gone into "safe mode," a state entered when the onboard computer has detected a problem. Multiple attempts were made to reinitiate contact with MGS, but none were successful. Mission planners went so far as to send a spacecraft-to-spacecraft signal from a European orbiter, Mars Express, in 2006, but to no avail. The mission of MGS was officially declared over in January 2007.

Despite the unscheduled conclusion of its endeavors, Mars Global Surveyor had been a resounding success. What began as a low-budget follow-up to the loss of Mars Observer turned out to be by far the most long-lived and productive journey made to the red planet yet.

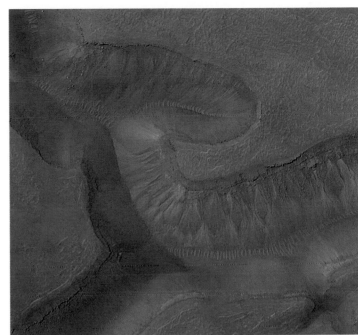

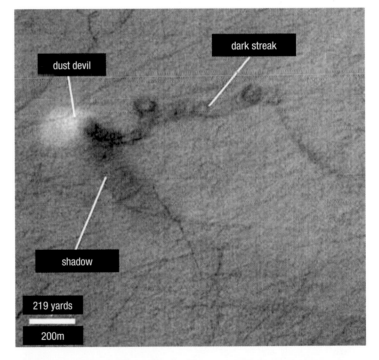

dust devil

dark streak

shadow

219 yards

200m

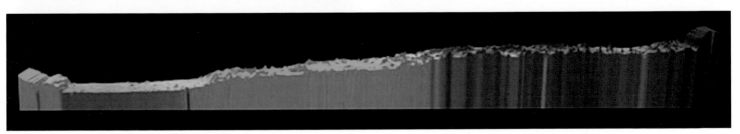

AN ODYSSEY TO THE GREAT GALACTIC GHOUL

Any member of JPL who works on the Mars program, and certainly most Russian planetary scientists and engineers, will tell you: Mars is hard. For decades the success ratio of all missions sent to Mars stood at only about fifty percent, with the number of flight failures skewed heavily toward the Soviet Union. But NASA had its fair share, too, and that bad luck reasserted itself in the late 1990s.

THE LOW-BUDGET AND VERY SUCCESSFUL Mars Pathfinder and Mars Global Surveyor missions had built confidence in the "less-is-more" school of thought, and it seemed that the bugs had been worked out of getting to the red planet on a budget. Two new missions were planned with the same philosophy in mind.

The first of these cost-conscious missions was the Mars Climate Orbiter (MCO). At a pre-launch cost of just under $200 million, it was in the same general price range as Pathfinder and MGS. MCO would specialize in monitoring the Martian climate, atmospheric environment, and incremental surface changes. At 745lb (338kg), it was a comparative lightweight, would be relatively inexpensive to launch, and would therefore please the accountants at NASA headquarters. The spacecraft was built for NASA by Lockheed Martin (as were most of NASA's Mars orbiters), and launched in December of 1998.

The second mission was the Mars Polar Lander (MPL), a 640-lb (290-kg) craft destined to explore the Martian south polar region. A primary goal was to search for water ice. To aid in this endeavor, MPL carried two "impactors," hard-landing modules that would detach before the lander set-down and hurl themselves into the Martian tundra. These robotic bullets would burrow up to 3ft (1m) below the sandy soil and measure the characteristics of the virgin materials below; by this means they would provide a first look at dirt that had not been wasted by the severe effects of the sun's

RIGHT: The ill-fated Mars Climate Orbiter (MCO) spacecraft during testing.

MARS CLIMATE ORBITER

MISSION TYPE: Mars Orbiter

LAUNCH DATE: December 11, 1998

LAUNCH VEHICLE: Delta II

ARRIVAL DATE: September 23, 1999

MISSION TERMINATION: September 23, 1999

MISSION DURATION: 286 days—mission failure

SPACECRAFT MASS: 745 lbs

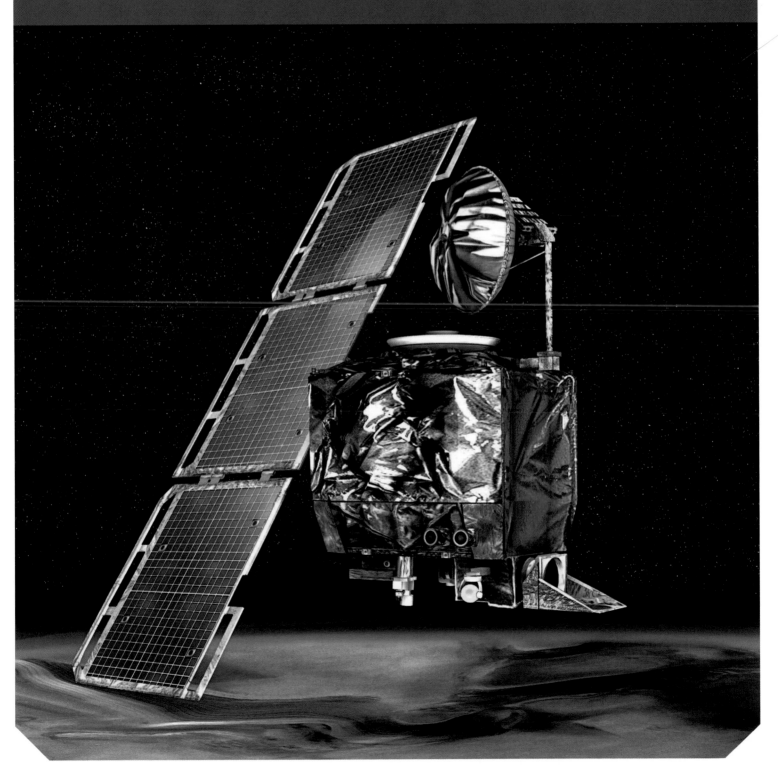

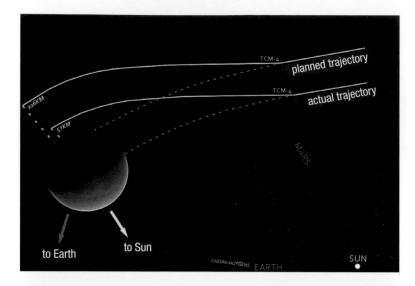

radiation and Martian surface chemistry and actions. MPL was launched in January 1999.

HOPES AND FEARS

Everything seemed to be going well during the first months, as the two spacecraft crossed the void to Mars. On September 3, 1999, MCO's trajectory was trimmed to put it in a proper configuration for aerobraking through the Martian atmosphere, a technique proved with MGS and now considered relatively routine. The braking rocket fired to slow MCO for aero-capture…

And then, nothing. Radio signals ceased in an instant, and concerned flight controllers scrambled to figure out if they had a temporary interruption, a loss of altitude control, or something more serious. A half-hour later, when MCO should have cleared the far side of Mars and reestablished communication, there was no signal. The mission was over in a heartbeat, and the engineering team immediately began searching for the cause of this failure.

Now all eyes were on the Mars Polar Lander. One failure per year was already alarming; two would be disastrous. Exactly three months later, on December 3, MPL was heading in to make its historic south polar landing. Like Pathfinder before it, this was a direct trajectory aimed at the surface; there would be no loitering in orbit before landing, a luxury only the Vikings, with their larger rockets and fuel supplies, had enjoyed.

The sixteen-minute engine burn that would brake MPL into the proper orientation for atmospheric entry started. Data from the spacecraft, as always delayed by the extreme distance between Earth and Mars, looked good. Everything was going well, as the craft plummeted into the thin Martian air.

And then, suddenly, everything went wrong. A half-hour later flight controllers were still trying to contact the spacecraft, which had last been heard from as it entered the atmosphere, headed for a region of Mars known as Planum Australe. Long after the lander should have signaled a successful touchdown, the radio was silent, and the mission was presumed lost. For the second time in three months, an investigative team was mustered to determine the cause of failure.

After months of digesting and evaluating data, two horrible, critical mistakes were found, one in each mission. It was apparent that the Mars Polar Lander had crashed when the onboard sensors and their software mistook the jolt of the landing legs unfolding and locking as an indicator that the spacecraft had set-down on the surface. The computer therefore dutifully shut down the descent engines to avoid polluting the Martian surface. Unfortunately, in actuality the probe was still well over 100ft (30m) in the air when

MARS POLAR LANDER

MISSION TYPE: Mars lander
LAUNCH DATE: January 3, 1999
LAUNCH VEHICLE: Delta II
ARRIVAL DATE: December 12, 1999
MISSION TERMINATION: December 12, 1999
MISSION DURATION: 334 days—mission failure
SPACECRAFT MASS: 640lb (290kg)

the landing rockets ceased firing, and the spacecraft accelerated into the rocky soil below. Scratch one lander.

The ultimate cause for the demise of the Mars Climate Orbiter was far more embarrassing. Somewhere in the long and sometimes tangled lines of communication between JPL and the contractor that built the orbiter, Lockheed Martin, the units of measure had been switched. Software supplied by the contractor for the operation of the spacecraft had been supplied in American units (feet, miles, inches), while JPL had been operating, as usual, in metric. A pre-orbital maneuver had placed the spacecraft into an erroneous trajectory, so instead of entering its aerobraking orbit of about 100 miles (160km), it had skimmed the planet at an altitude of about 70 miles (112km). Even the thin Martian atmosphere was too dense at that height for the spacecraft to survive, and it apparently broke up during its first aerobraking pass.

Both failures were embarrassing, but the press was particularly tough on the MCO debacle. NASA had lost two Mars missions within weeks of one another, and both in the same year. Was the "faster, cheaper, better" campaign to blame, with understaffed teams doing too much work and too few tests? Or was the contractor the culprit? Or perhaps the mission designs were faulty from the start? It was, in the end, a variety of factors that led to poor communication, insufficient oversight, and two lost missions.

THE HAND OF FATE?

But there were some who whispered, mostly in jest, about another culprit, one that had long haunted the linoleum halls of JPL and the Soviet/Russian space bureau. Whether black humor or grim reality,

there was really no question… the fabled Great Galactic Ghoul had struck again.

Depending on how you count them— whether an orbiter and lander are two spacecraft or only one, etc.—in total just over fifty spacecraft have headed off to Mars. Only about half have arrived successfully. It is a miserable success ratio, and while the figures are heavily biased toward the early days of robotic exploration of the solar system, Mars seems to eat spacecraft with distressing regularity. While it has been by far the most explored planet in the solar system after Earth, Mars has also been the hungriest in terms of spacecraft lost. By the mid-1960s, those laboring in Mars exploration programs wondered what made the planet just so hard to reach. There were plenty of reasonable explanations— distance, trajectories, the harsh space environment, and many more. But there was an almost surreal quality to the losses. True, they were heavily weighted toward Soviet spacecraft; yet,

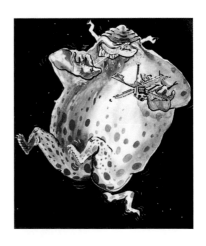

no matter how many were sent, to this day no Soviet or Russian spacecraft has been fully successful in operations at Mars.

So the engineers wondered… what exactly makes Mars so risky for robotic exploration?

A quirky explanation came from an engineer in NASA's ranks in the mid-1960s. While intended to be whimsical, his joke has survived to this day, and the name of the supposed culprit is invoked by mission planners with an uncomfortable smirk whenever another probe is lost. The evil menace responsible? The Great Galactic Ghoul.

The man who came up with this catchy moniker was John Casani. While a young engineer working on the Mariner program, Casani was interviewed by a journalist covering the space exploration beat. The Soviets had already lost five spacecraft in the region of the red planet, and the US was about to send its own first probes, Mariners 3 and 4, of which only one would ultimately survive. When questioned about the seemingly poor odds of success, Casani chuckled and invoked an evil space-being he called the Great Galactic Ghoul. The name has stuck ever since, immortalized in the occasional cartoon and mentioned with a nervous laugh by a new generation of Mars explorers.

But whimsy aside, *something* was happening to many Mars-bound spacecraft. A cause or likely explanation was always arrived at after careful examination of the evidence—spacecraft telemetry (the electronic signals sent to Earth), pre-flight assembly and testing records, and the like—but whatever the cause, NASA could not afford another failure after the twin losses of MCO and MPL came so close together. With the next launch window for a Mars mission nearing in 2001, another failure would likely shut down the program for years. The next mission simply had to succeed.

It was in this spirit of determination that the preparations for Mars Odyssey (MO) were completed. NASA returned to Lockheed Martin, the same company that had built the Mars Polar Lander, to build the spacecraft. Little needed to be said to the contractor's management about the stakes (though much certainly was), and the assembly of the spacecraft was held to tighter standards than ever before.

Mars Odyssey's specific mission was to use a suite of spectrometers and a thermal camera to search the surface for water ice as well as other traces of moisture. A radiation detector was onboard as well to study threats to future human voyagers to the red planet from high-energy exposure. And MO had one more role to fulfill—it was to be, along with MGS, a relay station for the upcoming Mars Exploration Rovers, scheduled to land in 2004. The mission was funded at about the same level as recent attempts, about $297 million. It was not a vast sum in spaceflight terms, but it was all that NASA was willing to gamble. As the saying goes in the space trade, failure was not an option.

MARS ODYSSEY

MO was launched on a now-standard Delta II rocket and swung into a polar orbit around Mars seven months later in October 2001. Once again, aerobraking was used to slow the spacecraft into a circular orbit from which it could pursue its goals. Its planned two-year mission began in earnest in February 2002.

When Jeffrey Plaut looks back on the early days of the Mars Odyssey mission, one discovery stands above the rest. "Finding and mapping ice in the soil of Mars was big. We made unequivocal observations and created maps of hydrogen in the subsurface of Mars down within the first couple of feet."

The gas was spotted by MO's Gamma Ray Spectrometer, which measures particles that result from cosmic rays impacting the Martian surface, including indicators of hydrogen. And where there is hydrogen in the soil, there's likely to be water.

"We saw that both polar regions, down to about 60 degrees north and south latitudes—what you might call the arctic of Mars—was just shot through with ice in the soil. That had been predicted, but there really was no way for anybody to make that measurement or map it until Odyssey came along."

The hydrogen-rich areas surveyed appeared to be between 20 and 50 percent water ice. "We ended up providing the landing target for the Mars Phoenix lander mission, which landed within Mars' 'arctic circle,' and it was able to scrape the surface with its robotic arm tools. This exposed water ice, and there was also ice exposed underneath the body of the lander—the soil was blown away by the landing rockets, and there were slabs of ice underneath it. That was a high point for me."

MARS ODYSSEY

MISSION TYPE: Mars orbiter

LAUNCH DATE: April 7, 2001

LAUNCH VEHICLE: Delta II

ARRIVAL DATE: October 21, 2001

MISSION TERMINATION: Continuing

MISSION DURATION: 15+ years

SPACECRAFT MASS: 829 lbs

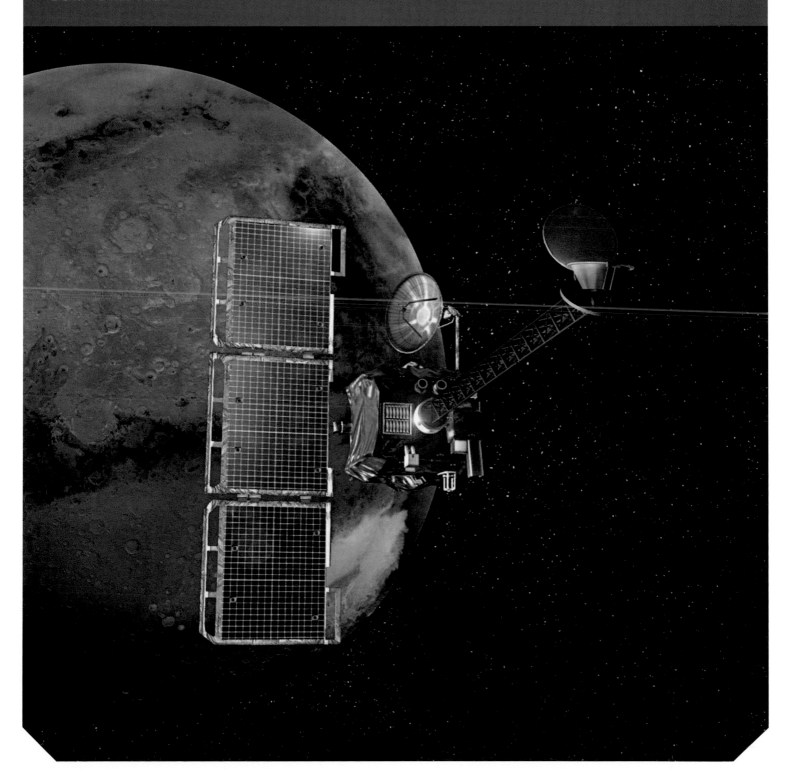

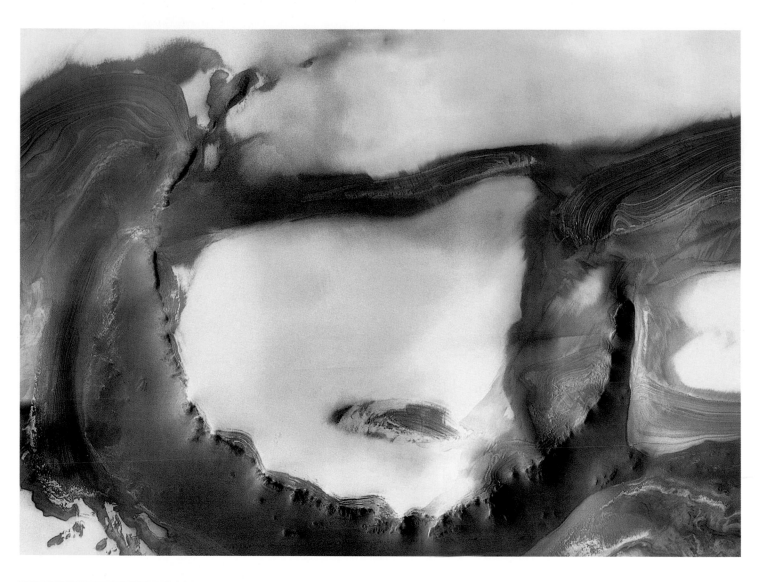

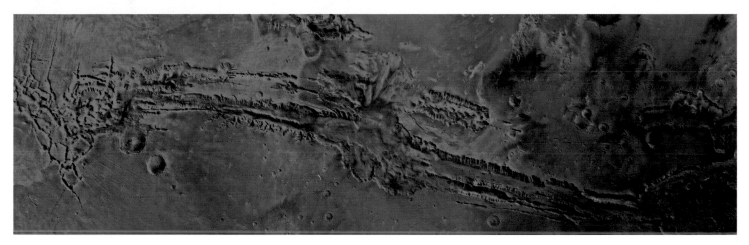

Everything went well until October 2003, when the radiation experiment ceased to function, probably due to electromagnetic damage to a circuit board. This was, however, near the end of the primary mission, and plenty of data had been gathered already. The rest of the spacecraft continued to function normally.

One of the most profound results to arise from the MO mission, indeed that which in large part it had been designed to fulfill, was the mapping of vast quantities of water ice distributed across Mars, especially near the poles. Virtually all of this is found sub-surface, existing as underground glaciers and ice deposits deep underground or near the surface of the poles. Indirect measurements via interpretation of water-associated mineral deposits refined the estimates. The question of where the thirsty planet had hidden much of the water long associated with the vast eroded features seen across the planet was, to a large extent, answered. Mission scientists concluded after long study of the data that there was enough water on today's Mars to cover the entire planet at a depth of about 30ft (10m) if it was all thawed and made its way to the surface. These findings pinpointed the landing site for a future mission, the Mars Phoenix polar lander.

After the primary mission ended in 2004, Mars Odyssey's assignment was extended... and then extended again. The spacecraft continues to circle Mars, sending back new data about the surface composition of the planet. MO also fulfills a secondary role as a relay station for the rovers below (it transmits data from the Mars Exploration Rover Opportunity regularly). MO is the longest-serving spacecraft ever sent to Mars, having entered its fifteenth year of service in 2016. And despite the failure in 2012 of one of its flywheels, used to orient the spacecraft, an onboard spare continues to allow MO to function in top form.

Mars Odyssey—the spacecraft that the Ghoul could never touch—is scheduled to continue its journey of exploration to at least the end of 2016 and very likely well beyond.

OPPOSITE TOP: Udzha Crater, as seen by the MO orbiter in 2010, is nearly buried in dust and ice. Measuring 28 miles (45km) in diameter, the crater is located near the north pole of the planet.

OPPOSITE BOTTOM: Chasma Boreale, a 1-mile/ 1.6km-deep canyon, as imaged by Mars Odyssey. The canyon is 350 miles (565km) long and cuts into the edge of the north polar ice cap.

ABOVE: Valles Marineris, the largest valley in the solar system, is seen here in images from Mars Odyssey (top) and the Viking orbiters (bottom). Note the vastly improved detail and color rendition from the newer, high-resolution cameras (top).

THE EXPRESS LANE: MARS EXPRESS

The next entrant into the Mars exploration arena would not come from NASA, nor would it be a Russian spacecraft. The European Space Agency, founded in 1975, had been launching Earth orbital missions and providing satellite launching services and scientific research for decades. Comprised of twenty-two member states, ESA flew its own astronauts with the Soviet Union in 1978, and joined the US-led International Space Station effort in 1983.

IN 1985, ESA LAUNCHED its highly successful Giotto cometary probe to Comet Halley, and dispatched a mission called Rosetta to comet 67P/Churyumov-Gerasimenko in 2004, making the difficult deep-space rendezvous in 2014 and ultimately landing a probe on the comet. Mars was a natural goal for the agency, and planning for such a mission began in the late 1990s. Mars Express (ME) was a two-part spacecraft: the Mars Express orbiter and a small lander called Beagle 2. The lander was built by the UK and its research goals were to analyze the surface adjacent to its landing zone and search for signs of life, past or present. Beagle 2's ingenious design incorporated the first "burrowing" mechanism sent to Mars. This device would be able to travel a short distance, worm its way into the soil to procure a small sample, and then be retracted into the lander so that the soil could be analyzed. The whole assembly weighed less than 73lb (33kg), and was about 2ft (60cm) wide before deploying its unfolding solar panels.

BROKEN BEAGLE

The Mars Express orbiter was a 1500-lb (680-kg) spacecraft that incorporated some of the instrumentation previously used on the failed Russian Mars 96 mission. In a sense, Mars Express would be the first successful Russian Mars mission, at least in terms of having some of its machinery flown there. The basic structure of the spacecraft was also repurposed for a Venus Express orbiter and the Rosetta comet rendezvous mission, both of which were highly successful.

The ME orbiter's goals included photo-geological studies via the use of a very high-resolution stereo camera. Mineralogical mapping and atmospheric analysis would be accomplished with a suite of infrared and ultraviolet spectrometers. Radar would map the Martian surface and deep into the soil, using two 60-ft (18-m) booms to act as a large, and extremely sensitive, antenna. Both ice and liquid water would be detectable down to 3 miles (5km) below the surface. Radiation sensors would investigate the interactions between the upper atmosphere and the solar wind, which is composed of charged particles streaming out from the distant sun. While the instruments may sound familiar, the design of ME was intended to perform in-depth investigations into some areas of Mars science that had not yet been attempted in such remarkable detail. Mars Express was a European multinational effort, so mission control was centrally located in

Darmstadt, Germany, and the mission was supported by NASA via JPL.

Mars Express launched on a Russian rocket in June 2003. After successfully attaining Earth orbit, the spacecraft was boosted into a rapid six-month Martian trajectory, arriving at the planet in late December. A few days before ME entered Martian orbit, it sent the Beagle 2 lander off on a separate course to enter the Martian atmosphere at an angle favorable to a landing, expected on December 24. As the orbiter prepared to enter Martian orbit on the 25th, controllers tracked the progress of the lander. It was bound for a region called Isidis Planitia, a flat plain that intersects the boundary of the Martian highlands, the ancient heavily cratered region, and the younger, smoother northern plains.

After separation on December 19, Beagle 2 entered the atmosphere on the morning of Christmas Day. Parachutes deployed to slow the lander, and were followed by rapidly inflated airbags designed to cushion the impact. But soon after atmospheric entry, contact was lost. Repeated attempts to reach the lander failed, and the $75 million craft was declared lost in February. Later images from Mars orbiters show Beagle 2, apparently in one piece, resting on the surface—it appears that the solar panels did not unfold properly, blocking its transmitter and denying the lander the power needed to operate.

OPPOSITE: The mission patch for the ESA's Mars Express.

ABOVE LEFT: Mars Express launches from the Baikonur Cosmodrome aboard a Russian Soyuz rocket on June 2, 2003.

ABOVE RIGHT: An illustration showing the water-finding radar in action. Both ice and liquid water would be detectable to 3 miles (5km) below the surface.

MARS EXPRESS TAKES CHARGE

In stark contrast, the ME orbiter fired its Russian–supplied upper-stage rocket engine for over a half-hour to brake into a successful Martian orbit. Once captured by the planet's gravitational field, subsequent firings rounded-out its trajectory into the desired elliptical 160 × 7100-mile (257 × 11,426km) orbit. While the orbiter was designed to be capable of aerobraking, it was not deemed necessary—with over 1300lb (589kg) of fuel, of which just over 800lb (362kg) should be needed for the primary mission, there was plenty to spare.

The instruments began their work almost immediately. In May 2004, the first antenna boom was extended to enable the subsurface radar readings. It was a slow procedure, and carried

its own risks—there was concern about a whiplash seffect when the long boom was extended. At first it did not lock into place, but after a bit of time exposed to sunlight, the locking mechanism warmed and engaged. The second boom was extended without mishap in the following month. This long-baseline radar would return some of the most impressive data gathered during the long mission to follow.

Discoveries began to come in almost immediately. The first was a positive detection of water ice at the Martian south pole—long suspected, and now confirmed. Two months later, in March, the detection of methane in the atmosphere was announced. It was a small amount, but enough to raise the question of whether or not the origins of the gas might be biological in nature, as methane can be a byproduct of microbial metabolism. Intriguingly, since methane is rapidly removed from the Martian atmosphere, this detection suggested that it must be periodically replenished in some way. Whether this was via biology or by a non-biological mechanism was open to question. Soon thereafter a similar result was announced regarding ammonia, with similar implications—a possible biological origin. Again, the detection raised more questions than it answered, but that is how planetary science works.

YEARS OF SUCCESS

In the following years, Mars Express compiled an impressive list of discoveries and observations. Hydrated minerals were found on the surface, once again confirming the role of water in the planet's environment. The ability to map these concentrations in detail also gave scientists a better understanding of geological and hydrological processes active in the formation of the planet's terrain. The orbiter also made the closest flyby of the Martian moon Phobos to date.

ME has continued to send home data since its commissioning. ESA has coordinated various investigations with NASA, including an attempt to visually locate the MGS orbiter when it fell silent (it was not located, despite their best efforts). There have been challenges—for example, as with so many spacecraft operating in distant locales. Radiation began to take its toll, limiting the functionality of some of ME's computing capabilities. However, with numerous mission extensions, ME continues to investigate the mysteries of the red planet to this day, and with its enormous fuel reserve, is expected to continue in service well into the 2020s.

ABOVE: The Mars Express orbiter weighed almost 2500lb (1130kg) with fuel for the flight. While the spacecraft was designed to be capable of aerobraking, the fuel allotment was so generous that it used its braking rockets to settle into orbit around Mars.

OPPOSITE TOP: The Promethei Planum region as imaged by Mars Express. The area is just 14 degrees from the south pole of the planet, and has ice depths of over 2 miles (3km) in some places.

OPPOSITE BOTTOM: Mars Express imaged this region, Ladon Valles, in 2012. The crater above left is 273 miles (440km) wide. Craters within the larger basin are interconnected by interesting fracture patterns.

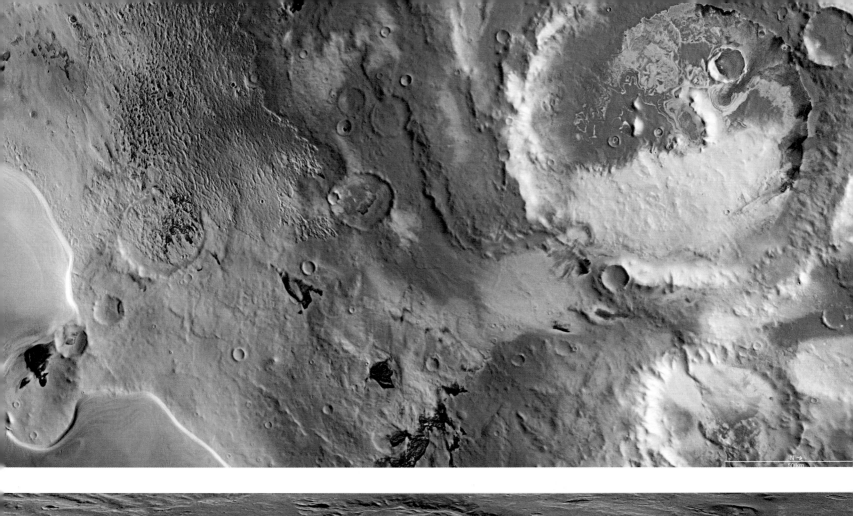

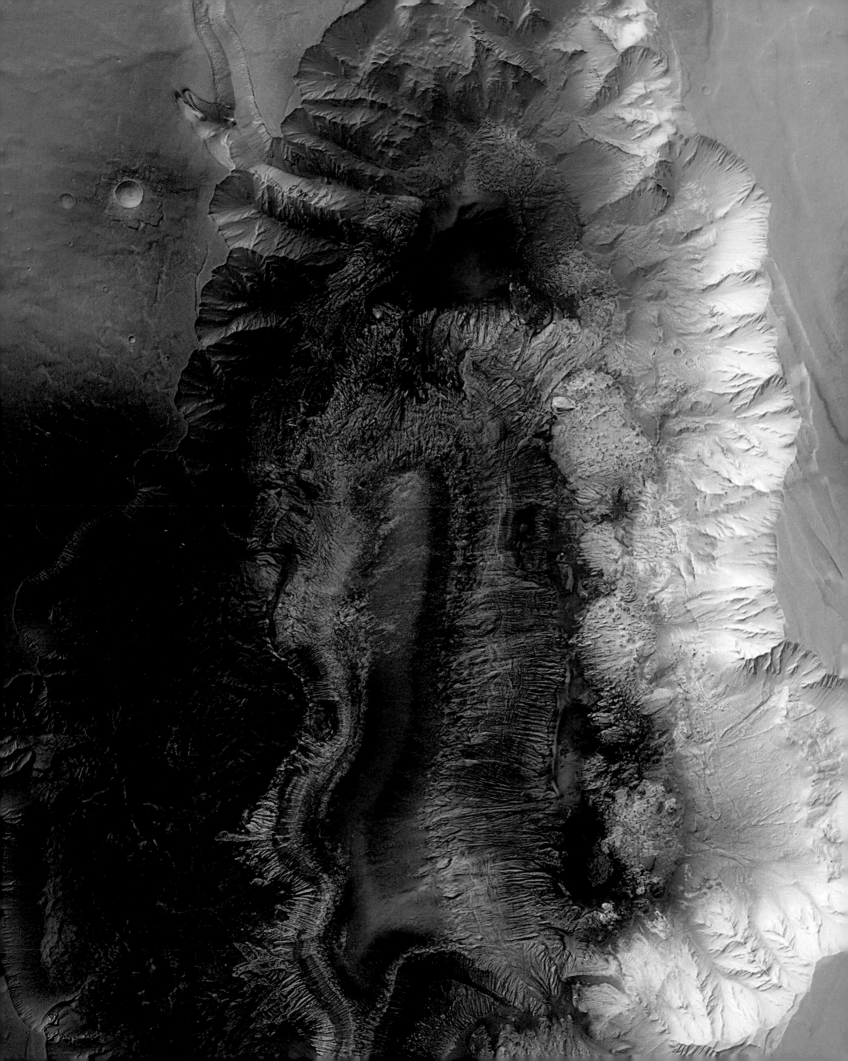

MARS EXPRESS

MISSION TYPE: Mars orbiter
LAUNCH DATE: June 2, 2003
LAUNCH VEHICLE: Soyuz FG-Fregat
ARRIVAL DATE: December 25, 2003
MISSION TERMINATION: Continuing
MISSION DURATION: 13+ years
SPACECRAFT MASS: 1468lb (665kg)

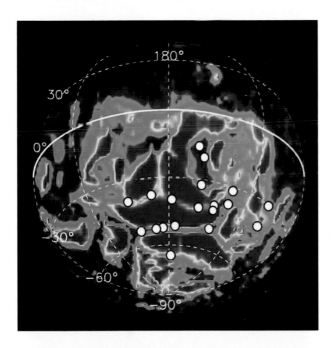

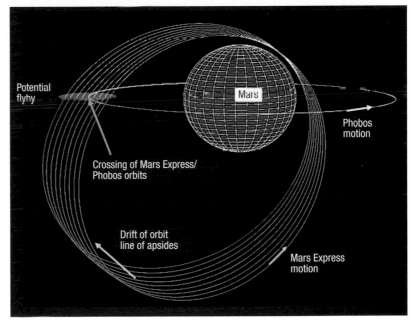

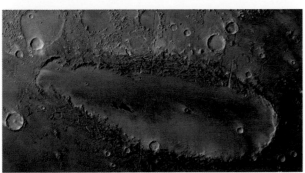

OPPOSITE: Hebes Chasma, just north of the enormous Valles Marineris canyon, is seen here. The mesa in the center seems to have undergone a partial collapse (square region at the bottom).

ABOVE LEFT: A map of Martian auroras, seen by Mars Express in ultraviolet. This map of auroras, modeled on a decade of measurements, helps define the extent and strength of Mars' magnetic field.

ABOVE RIGHT: This trajectory plot shows Mars Express' closest pass to the Martian moon Phobos. Swinging by the planet at an altitude of just 33 miles (52km) allowed for unprecedented observations of the tiny, irregular moon.

LEFT: Orcus Patera, seen here in 2005, has been an object of interest since the early Mars missions. About 23 miles (38km) long, It is just west of Olympus Mons. Scientists are still trying to determine what caused its curious, elongated shape.

FOLLOW THE WATER

NASA's Mars exploration program blossomed in the first decade of the twenty-first century. If Pathfinder and the Mars Global Surveyor were the primer after twenty years of neglect, this was the explosion. Within just a few years, three orbiters were criss-crossing the skies of the red planet every few hours. MGS, Mars Odyssey, and Mars Express were all functioning in top form, providing plenty of bandwidth for the relaying of surface operations data and imagery, as well as more detailed information about the terrain below than any previous landing attempt had enjoyed.

THE AMBITIOUS MARS EXPLORATION ROVERS (MER) program was the first to take advantage of this deliberately designed communications infrastructure. MER built on lessons learned from the Pathfinder mission, utilizing the same entry and descent trajectory and airbag landing method. The twin rovers, named Spirit and Opportunity, were launched in June and July of 2003, respectively, on a trajectory aimed squarely at the surface of Mars. Like Pathfinder, there was no intention to orbit the planet, but would head directly toward the surface.

The story of Spirit and Opportunity is a convoluted one. At the dawn of the twenty-first century, NASA's Mars program had two colossal black eyes—the failed Mars Climate Orbiter and Mars Polar Lander missions. Mars Odyssey was soon to launch, but in 2000 its success was far from assured. And there was another launch window—a very favorable one—in 2003, when Mars would be closer to Earth than it would be again until 2087. Such proximity meant that JPL would be able to get more machinery to the planet with the rocket boosters available to them, and hence more science for less cost. NASA naturally wanted to take maximum advantage of the occasion.

Harkening back to the success of Pathfinder, planners came up with a lander design based on the Mars Polar Lander. Although the MPL mission had failed, it was an immediately available and tested design. However, a single, static lander seemed insufficiently robust, given the unique launch window. The plan morphed instead

into a giant orbiter, to be packed with instruments that had been designed but not yet flown, due to previous mission priorities and budget limitations. However, as the idea was developed, it became clear that three years was simply not enough time in which to build such a complex machine. NASA had decisions to make, and quickly.

ATHENA IS ADAPTED

A JPL engineer named Mark Adler came forward with a plan to build on the successes of Pathfinder, adding an instrument package that had already been developed over many years in association with Cornell University. The experiments were collectively called Athena, and the landing mission for which they had been intended had never flown. Why not put them into a rover? The beauty of the plan was that much of the long, hard work involved in designing the complex experimental package had already been done, and the Pathfinder mission had proved JPL's ability to build a successful rover.

The plan was approved very quickly, and work began on the Mars Exploration Rovers. The team included a number of familiar faces from the Pathfinder crew, but some of JPL's old hands—

ABOVE: A mission patch for the Mars Exploration Rovers program.

OPPOSITE TOP: Spirit landed in Gusev Crater, near the center, and Opportunity on roughly the opposite side of Mars in Miridiani Planum, to the right. The geology of the landing zones could not have been more different.

OPPOSITE BOTTOM: Technicians check over Spirit before it is loaded atop the rocket at the Kennedy Space Center.

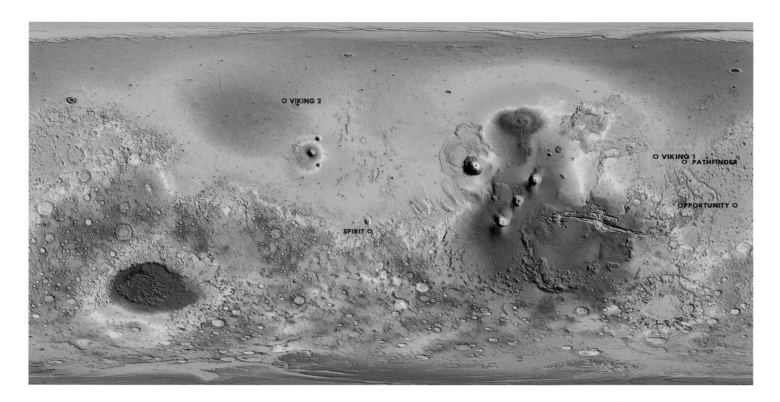

veterans of the Viking mission—were brought into the mix as well. As the engineers sketched designs and crunched numbers, one thing became rapidly apparent: the MER rovers were going to be larger and heavier than Pathfinder, so the parachute, rockets, and airbag systems would have to be larger as well, in order to support the additional weight. They would once again be pushing the limits of what was known about landing on Mars. As MER's design evolved, the spacecraft gained more weight, adding to the challenge. This was not a huge surprise—they had been through such engineering efforts before—but time was short. The launch could not be allowed to slip from the 2003 window and the workload quickly assumed Herculean proportions.

Testing of the landing system components took center stage. A larger and more robust parachute was needed, harkening back to the days of Viking. But with the direct-entry trajectory of MER, the spacecraft would encounter the Martian atmosphere somewhat faster than the Viking landers had, so existing designs had to be questioned and tested. As this process progressed, the MER parachutes ripped and snarled in wind tunnel tests, challenging the engineering teams to their limits.

The airbag design challenged them as well. The landing system for Pathfinder had worked brilliantly, but simply scaling up that design did not mean it would work for MER. The Pathfinder spacecraft—lander and rover—had weighed in at just over 600lb (272kg). MER was closer to 1200lb (544kg), and that meant that every aspect of the landing system had to be rigorously tested in order to accommodate the added mass.

The rovers were almost a treat to design in comparison. Pathfinder's Sojourner rover provided some reference points. Sojourner weighed only about 30lb (14kg) and the MERs would be well over ten times that, but the basic configuration scaled well. The same suspension system was used—a unique contraption named "rocker-bogie." As before, six wheels were mounted, three on each side, on the uniquely articulated swing-arm system. Sojourner had shown that a rover could climb far steeper and more complex terrain than would have been possible with a more traditional suspension design. One new design consideration

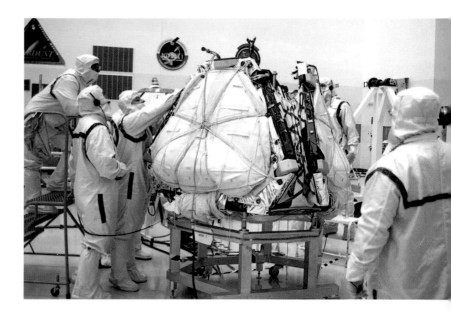

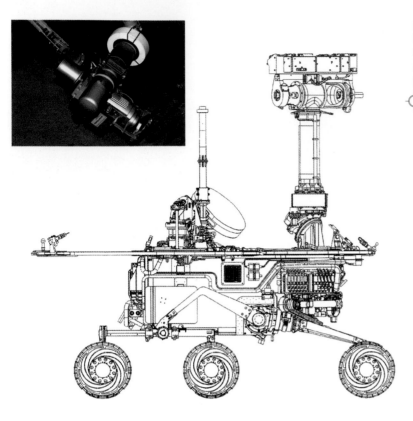

Pathfinder's, and could provide images in 3D when desired. A set of low-resolution navigation cameras, with a wider field-of-view to allow the rover drivers to see a broader swathe of terrain, were added to the 4-ft (1.2-m) mast. These "NavCams" were a must-have for intelligent route planning. There were an additional four smaller cameras mounted on each corner of the rover body, designed to provide hazard avoidance.

The rest of Athena's instrumentation was mounted on the end of a robotic arm that extended from the front of the rover. An infrared spectrometer would allow for close-up investigation of the mineralogy of rocks and soil. An alpha-particle X-ray spectrometer, an improved design of Sojourner's APXS, would evaluate the chemical content of surface targets. A microscopic camera was also mounted on the arm, as was a set of magnets for investigating ferrous dirt samples. Finally, a Rock Abrasion Tool (RAT), which incorporated a rotating wire brush, would be capable of removing dust and grinding-down the face of rocks, providing a clean surface for evaluation. Pathfinder's APXS had to peer through surface dust when it investigated rocks, resulting in a mixture of signals that could be potentially misleading. The ability to brush off and even grind down the surface of a rock target prior to an APXS reading was an important advance.

was incorporated—the front wheels were steerable. Sojourner had steered like a bulldozer, with the wheels on one side locking while the other side drove the rover into a turn. Steering that way wasted a lot of energy. While steerable wheels added complexity, the design allowed for more efficient and precise driving across the undulating and rugged Martian surface.

Designers once again chose solar panels to power the rovers. They had considered a nuclear power source, as had been used on the Viking landers, but solar panels had worked well on Pathfinder and were an affordable, workable choice. To provide the extra power that the larger MER rovers would need, there were "wings" hinged to the topside of the rover that would unfold after landing, giving it somewhat the appearance of a beetle. These would roughly double the area of the solar panels and provide much more power to the rover. This increased area of exposure would be especially critical if the top of the rover collected a covering of Martian dust, as was expected.

Of course, the MER rovers were merely machines designed to provide mobility to the science instrumentation. Mars missions are all about the science. The Athena science package evolved to capitalize on the mobility provided by a rover. The cameras were mounted on a mast, providing a far better view of the terrain ahead than if they were mounted at deck-level. They incorporated a set of stereo optics operating in high-resolution, three times that of

MARS EXPLORATION ROVERS

MISSION TYPE: Mars rovers
LAUNCH DATE: Spirit: June 10, 2003, Opportunity: July 7, 2003
LAUNCH VEHICLE: Delta II
ARRIVAL DATE: Spirit: January 4, 2004, Opportunity:
January 25, 2004
MISSION TERMINATION: Spirit: March 22, 2010,
Opportunity: Continuing
MISSION DURATION: Spirit: 6 years, 2 months, 10 days,
Opportunity: 12+ years
SPACECRAFT MASS: Lander: 767, rover: 390lb (176kg)

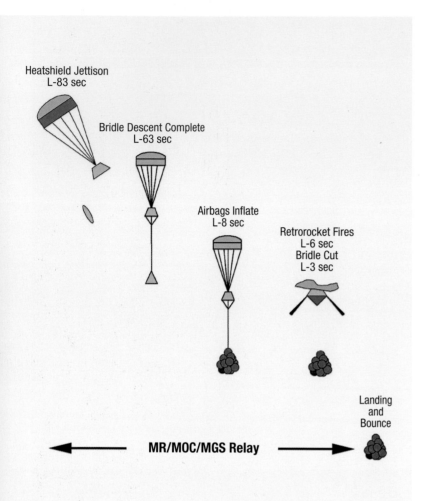

Heatshield Jettison
L-83 sec

Bridle Descent Complete
L-63 sec

Airbags Inflate
L-8 sec

Retrorocket Fires
L-6 sec
Bridle Cut
L-3 sec

Landing
and
Bounce

← MR/MOC/MGS Relay →

SPIRIT AND OPPORTUNITY BEGIN THEIR JOURNEY

Spirit began its journey to Mars on June 10, 2003, and Opportunity on July 8, 2003. Both traveled just over six months to get to their common destination.

On January 4, 2004, Spirit was the first to slam into the Martian atmosphere, traveling at over 12,000mph (19,312km/h). As with previous Mars landers, at this point the spacecraft was completely on its own. Due to the long delay of radio signals crossing between Earth and Mars, the spacecraft had to be capable of "thinking" for itself. Spirit knew its position in space, thanks to its inertial guidance system, and it knew the location of the preferred landing zone on Mars. For the next six minutes, the spacecraft would be responsible for making its own way through the fiery, violent Entry, Descent and Landing sequence (EDL) to get from space to the surface of Mars.

Flight controllers at JPL hunched over their consoles. There was nothing to be done other than watch, but each person was transfixed by the drama playing out so many millions of miles away. While the controllers could not talk to Spirit, they were able to monitor the descent thanks to the "heartbeat" tones it sent back to Earth—simple signals that could penetrate the ionized fireball created as Spirit descended. Specific tones were used to communicate various status updates, such as parachute deployment and opening, deceleration rates and maneuvering adjustments.

Within moments the heat shield was heated to 2,600°F (1,426°C) as it sliced through the thin, frigid Martian air. Though the Martian atmosphere is incredibly thin, it is still dense enough to create a lot of friction and heat. After four minutes Spirit had slowed to about 1000mph (1609km/h) and the parachute was shot out of its mortar, opening in a supersonic airstream about 30,000ft (9144m) above the surface. A minute later, the heat shield dropped free and the lander was winched down, away from the protective aeroshell in which it rode, via a tether. Within seconds it was dangling from a 65-ft (19.8-m) cord.

At 8000ft (2438m), onboard radar fired up to provide rate of descent and transverse motion data to the computer, allowing it to calculate how long the retrorockets would need to slow the craft to the proper rate of descent and arrest any sideways motion.

Concurrently, an intricate orbital ballet was occurring overhead. Planners had timed the landing to coincide with a pass of the Mars Global Surveyor orbiter, allowing it to relay signals from Spirit to Earth. A decade of careful mission planning was now paying dividends—it was the first time an orbital relay link had been available to make communicating with a lander easier and more reliable.

As they had with Viking before it, the MER mission planners picked two landing sites on opposite sides of the planet. The process was long and arduous, but the overriding criterion was simple: follow the water. Using the best images the Mars Global Surveyor could provide, the landing team debated one site after another. After much argument and evaluation, Spirit would be sent to a region known as Gusev Crater. Its wall had a break in it, through which it was thought that water might have once flowed from a neighboring region, possibly depositing sediment into the crater floor. Spirit targeted the center of the crater.

For Opportunity, the team selected an area called *Meridiani Planum*, the Meridian Plain, on roughly the opposite side of the planet. Both sites were roughly equatorial, to optimize the use of solar panels that powered the MER rovers. Meridiani had been examined from orbit, and appeared to have vast deposits of hematite, a mineral formed with the interaction of water. This finding indicated that the area would be a promising site for investigating the story of ancient water on Mars.

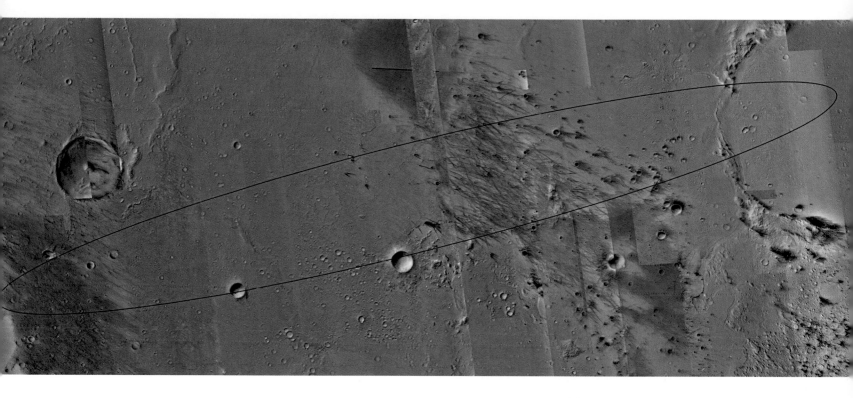

The Pathfinder-tested airbags inflated in an instant, and the retrorockets fired to slow Spirit to a crawl just 40ft (12m) from the ground. The tether was cut, and in three seconds the spacecraft made its first bouncing impact on Mars. The scientists wanted Spirit to land somewhere within a 54-mile (87–km) long oval called the landing ellipse; the engineers just wanted it down safely. As it turned out, both parties had their wishes granted.

The bulbous package hit the ground and bounced again and again, just as Pathfinder had done, rolling to a stop just over 6 miles (10km) from the center of the landing zone. After a 300-plus million mile (483 million km) journey, that was nearly a bullseye.

Over the next three hours, protective side-panels were unfolded, righting the lander, and the landing bags were retracted via winches and cables. Spirit was down and safe in Gusev Crater, and within days NASA's remarkable decade of wheeled exploration on Mars would begin.

OPPOSITE: Diagram showing the final phases of the Entry, Descent and Landing (EDL) phases of each MER rover. Before the first image to upper left, the spacecraft will have entered the atmosphere at 12,000mph (19,300km/h), slowing to just under 1000mph (1600km/h) 4.5 minutes later, when it jettisons the heat shield.

ABOVE: Spirit's landing zone, or ellipse, was 48 miles (78 km) long by 6.5 miles (10.4 km) wide. It finally rolled to a stop not far from dead center, quite a feat given the ferocious winds it experienced during landing.

RIGHT: Rob Manning, Chief Engineer for the MER mission, fist pumps upon hearing news of a successful touchdown.

BLUEBERRIES, DUST DEVILS, AND OTHER MARTIAN DELIGHTS

After arriving at Gusev Crater, it took Spirit almost eleven days to actually get roving. And though events moved with often terrifying speed while actually *landing* on Mars, once there, caution was the watchword. With no means of fixing robotic machines on far-flung worlds, prudence suggested proceeding with care.

Ever so slowly, Spirit rolled off the lander, using one of the unfolded metal petals as a ramp. After a pause on the surface, the rover's cameras imaged its surroundings from ground level. The landing site looked different from anything encountered by either the Viking landers or Pathfinder. The floor of Gusev Crater was flat and wide, with smaller rocks in abundance. It looked perfect for driving and there were plenty of small targets to investigate. Panoramic imaging was done to determine the best course forward, and the landing site was dubbed "Columbia Memorial Station," in honor of the crew of the space shuttle Columbia, which had been lost during re-entry the year before.

VISUAL TREATS, GLITCHES, AND A SUCCESSFUL LANDING

For the first couple of weeks, Spirit moved slowly across the surface, gathering data from its surroundings and testing driving techniques and mobility. The visual imagery that flowed in to mission control was fantastic. These were a huge improvement over those received from Pathfinder, and the huge panoramas were pounced upon by the geologists, who studied each image with relish. Locations of the crater floor were selected for examination by the instruments onboard the rover, and it was at this point that the first disappointment of the mission emerged: early readings showed that Gusev was covered in ancient lava flows; the materials there were basalts, or volcanic rock. If it had once been drenched in water, as the orbital pictures had seemed to indicate—the crater had appeared from orbit to be a giant lake fed by a breach in its wall—volcanic activity had subsequently buried the evidence. There was still much to be learned, but the search for an ancient watery environment would have to wait.

Then, the first gremlin struck the rover on Martian day (or sol) number 17, when Spirit abruptly stopped communicating. Commands were sent Marsward from mission control in an attempt to diagnose the problem. The following day, the rover sent back a brief message—a single beep—to acknowledge that it had received a message, but that it had switched into fault mode. (This is not dissimilar from your home computer rebooting into "safe mode," when the computer restarts, realizes it is having a problem and then resets itself to a basic configuration.) The engineers did not know if this represented a simple bit of software corruption, or a more serious hardware problem.

As this problem was being worked on, above Mars, Opportunity continued its unstoppable rush toward its own landing. At this point, the flight teams had to manage one ailing rover even as the second machine was about to land. It was a harrowing time.

OPPOSITE TOP: Spirit looks back at its landing stage shortly after rolling onto the surface. Note how the landing bags, which would have spread out around the lander after touchdown, have been retracted by cables and winches to make sure they did not interfere with the descent.

OPPOSITE BOTTOM: The first full image form Spirit showed the bleak surface of Gusev Crater. While it had been hoped that the area would reveal water-altered features, it turned out to be a relatively plain basaltic plain.

ABOVE: Spirit as seen on the plains at Gusev. The surface is a composite of surface images, with a digital model of the rover added.

Two days before Opportunity's landing date, Spirit relayed a message back to JPL via the Mars Odyssey orbiter, indicating that it was not entering the required "sleep" mode during the Martian night, when the solar panels were not generating energy. It was idling all night, wasting battery power, and even overheating the electronics. A thorough look at the engineering data that was sent back led JPL's programmers to rewrite the software commands stored on the rover's flash drives.

As this drama was unfolding, Opportunity followed a now familiar routine on the other side of the planet, successfully landing within its own target ellipse in Meridiani Planum on January 25. Once the cameras were active, the mission scientists realized something remarkable: if Spirit had landed in a golfer's sand trap, Opportunity had made a hole-in-one. The lander had rolled to a stop inside a depression named Eagle Crater, and the first views of the area indicated layers in the crater wall—possible signs of sedimentary activity. And while it was too early to know for certain, those layers might have been deposited by water sometime in the distant past. The sting of Gusev's bland volcanic plain was mediated by the discovery of geological riches.

Eagle Crater was a hole dug out of the ancient Martian topography by a violent meteoritic impact, offering a stacked view of millions of years of geological evidence like a huge, drilled core-sample. As the chief scientist on the mission, Steve Squyres, enthused: "We lucked out… we discovered that we were in a giant impact crater that had all the things that we could have wanted, exposed there in the wall of the crater. In two months much of the important science was revealed to us."

However, that was just the beginning of the discoveries at Eagle Crater. Within days of rolling off the lander, Opportunity spotted little blue-gray balls all over the crater floor—they looked like BBs—

that took everyone by surprise. Were these glass balls formed by the explosive impact that created the crater? Or were they beads of volcanic glass—known on Earth as lapillae, that were like glassy hailstones resulting from volcanic eruptions? Closer examination revealed a far more intriguing explanation: these spherules—and there were thousands of them—were composed of hematite, a mineral formed by interaction with water. Bingo.

As Opportunity got closer to the crater wall, the amazing images just kept coming. Besides the obvious sedimentary layering that was becoming clearer and clearer, more spherules were found there, embedded in the rock face. As the softer material around them eroded away, the spherules would roll out and collect on the surrounding floor. This strengthened the formed-in-water hypothesis about the spherules, now nicknamed "blueberries."

The wall of the crater was closely examined. It turned out to be packed with a sulphate mineral known as jarosite, which forms via the evaporation of briny bodies of water.

Then, the eagle-eyed sedimentary geologist on the team, John Grotzinger (later to be the mission scientist for the Curiosity rover mission) spotted something… there was an exciting element to the layering. Known to geologists as "cross-bedding," this pattern indicated the presence, at some time in the ancient past, of flowing water. The close-up examination of Mars was revealing long-held secrets of its watery past.

ROVING ACROSS MARS

After a two-month investigation of its landing site, Opportunity headed off for its first long drive. The MER rovers generally moved at a rate of less than 2ft (60cm) per minute, so there was plenty of time to visually explore the surrounding terrain. In May 2004, four months after landing, the rover reached its first major destination,

ABOVE: Opportunity's second stop, once it left its landing site at Eagle Crater, was the 426ft (129m) Endurance Crater. This was the first major stop in a trek that would last well over a decade, and continues.

BELOW LEFT: This image from the Mars Global Surveyor orbiter shows hematite scattered across the surface. Concentrations range from 5% (blue) to 25% (red). This finding made Meridiani Planum an attractive location for a landing.

BELOW RIGHT: Hematite spherules, called blueberries, were found at Eagle Crater and elsewhere across Meridiani Planum. Thesse water-formed inclusions are left behind when softer soil erodes away.

Endurance Crater. After a careful reconnaissance of the 66ft/20m-deep depression's rim, the rover descended to the interior via the most promising-looking path in mid-June for a six-month stay.

Endurance Crater provided another set of rock layers to study, right at the entrance. As on Earth, the layers followed the basic rules of sedimentation, with younger layers overlying older ones. In the bottom of the crater were sand dunes, but the planners decided to avoid the potential sand trap for the moment, investigating them from a safe distance. They did test the rocks found there, which showed that they had been altered by water after they rolled into the crater. Endurance had apparently at one time held a lake of liquid water for a long period of time.

0 %   20 %

This was an important discovery on a world long presumed to be bone dry.

After six months at Endurance, Opportunity departed the crater. En route to its next goal, it visited its own heat shield, examining the effects of the impact of the metal disk on the Martian surface—analyzing a point of impact by an object of known mass could reveal important data about the hardness of the local surface. The rover then headed off to distant Victoria Crater, a magnificently carved, 2,400ft/732m-wide formation that looked interesting from orbit, with eroded "fingers" all round the rim that dug far back into the surrounding terrain. Along the way, the first meteorite on Mars was spotted and investigated, dubbed Heat Shield Rock, after its proximity to the previous stop. The composition of the basketball-sized impactite was found to be essentially the same as those found on Earth. After this short stop, Opportunity drove on.

In the meantime, Spirit had largely resolved its "mental" issues and moved on. A month after landing, it used the RAT rock-grinding tool for the first time in the mission. A rock named Adirondack was the target, and the 2in/5cm-wide abrasion of the surface allowed for a look at what lay within the rock. Using the microscopic imager, geologists were delighted by crystalline structures found within, even though it was merely another type of volcanic basalt. But the RAT had performed splendidly.

Spirit headed off to a region dubbed Columbia Hills, also in honor of the lost space shuttle. Along the way the engineers noticed the power beginning to wane—there was a buildup of dust on the solar panels. Since the primary mission was initially pegged at ninety days, this had been considered an acceptable risk when deciding between nuclear or solar power for the MER rovers. A few engineers might now have reconsidered the choice of a nuclear power supply, but that would have to wait for the next rover mission.

Eventually, a dust devil came along and provided a reasonably thorough cleaning of the solar panels, allowing both rovers to

continue their missions, despite ongoing coatings of fine-grained dirt. These mini-cyclones had been observed before, but nobody could have said with assurance that they would be sufficiently frequent, or powerful (Martian air is very thin) to perform as a Martian car wash. But in the end, the dust devils kept the rovers roving. These period air-brushings were able to boost power collected from the distant sun by almost as much as fifty percent.

By mid-2005, Spirit was making the first Martian hill-climb in history, slowly ascending Husband Hill. The rovers were driving mostly autonomously at this point, the drivers having turned on this feature once it had been tested earlier in the mission. Each day, route-planning teams would evaluate the previous day's drive (if any), look at the terrain ahead, note any potential obstacles or dangers, and give the rover general parameters on how to proceed. The machine was then capable of making its own limited

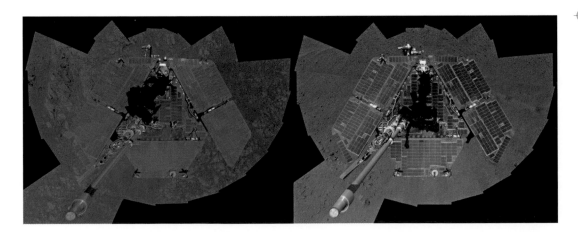

ABOVE: A circle ground by the RAT tool on the rock Adirondak. The grinding provided a much cleaner surface to allow for more accurate readings from the Alpha Particle X-ray Spectrometer.

LEFT: Two self-portraits taken by Opportunity in January 2014 (left) and March 2014 (right). During the interim, windy conditions cleaned dust off the solar panels, increasing power output by 70%.

OPPOSITE: In 2007, Spirit's wheels scuffed-up some soil, showing the silica-rich white soil below. This was yet another indication of the past presence of water.

marvel, and its transmitting bandwidth made MGS seem like a dial-up modem in contrast to MRO's broadband. The MER rovers would be well-supported to transmit whatever data they could collect for the remainder of their surveys.

Spirit continued its slow progress toward an overwintering station, despite slippery inclines, the occasional sand trap, and rocks jamming in front of the wheels.

Plans for both rovers' winter layover included atmospheric studies, examinations of nearby rocks, and extended observation of the weathering of the rover's tracks. Any chance to observe the effects of the Martian weather on freshly disturbed soil would aid in understanding the makeup of the soil and its interaction with Martian weather. With the rover parked for a period of time, geologists could, in effect, create time-lapse images of these changes.

By mid-2006, Spirit was deep into the Martian winter, with energy gathered from the solar panels down to about one third of the optimum amount. Each day (sol) generated only about enough power to run a 100-watt light bulb for around three hours, so the energy allowances were tight. Adding to the rough conditions, temperatures plunged to -143°F (-97°C). Despite these challenges, the rover continued its studies, adding a huge, high-density panoramic photo-survey to the workload.

As winter retreated and Spirit began driving anew, the ailing front wheel delivered an unexpected and marvelous tidbit of scientific discovery. As the wheel dragged along behind the rover (which was continuing to drive in reverse), it scuffed the red soil and revealed white material underneath. This was a surprise to everyone, and merited a closer look. Detailed examination showed that the white dusty material was rich in silica and appeared to represent an ancient watery environment, ideal for microbial life. On Earth, these kinds of conditions are traditionally seen near hot springs—perfect conditions for microbial colonies.

As Spirit moved on to new terrain, intense dust storms began once again to hamper the mission. By mid-2007, both rovers were being affected, and the dust was so thick in the atmosphere that the power was down to a tiny fraction of what was needed to continue—there was a real risk of losing one or both rovers due to battery failure. The machines simply could not exist on such low energy rations for much longer, with energy reserves barely equivalent to what it would take to run your refrigerator's light bulb for an afternoon. Dusty skies persisted into 2008, and much of that year saw Spirit operating in minimum conditions. Toward the end of the year, Spirit started short drives again, in part to enhance the tilt of its solar panels in an effort to improve energy generation.

set of decisions, stopping, and phoning home if it encountered unexpected difficulties. This bit of artificial intelligence embedded in the rover's computer avoided, for the most part, the long half-hour wait for a round-trip radio signal to mission control, which made traditional joy-sticking impossible.

When Spirit crested the hill, a search was begun for a suitable spot to park for the encroaching Martian winter. With the already faint sun dipping ever lower in the sky, power would be limited. The mission scientists were already shutting the rovers off during the cold Martian nights to conserve power. Adding to the seasonal challenge was occasional sky darkening due to dust storms; energy was at a premium. And mechanical problems were coming into play: one of Spirit's wheels, the right front, was experiencing difficulty, and there was a short-circuit somewhere in the electrical system that was draining energy. Spirit's mission, while productive, seemed to be cursed by small anomalies.

Spirit was now driving backward in order to lessen the strain on its ailing wheel. Diagnostic efforts using Spirit's clone at JPL's Mars test-bed were inconclusive—they would simply have to monitor the wheel throughout the remainder of the mission.

OVERCOMING PROBLEMS

Communicating with the rovers always required some choreography of the available assets orbiting Mars, but when the Mars Odyssey orbiter began experiencing intermittent difficulties, this became even more complex. But in a case of perfect timing, NASA's newest addition to the Mars team, the Mars Reconnaissance Orbiter, NASA's newest, arrived in March 2006. MRO was a technological

degrading, resetting more often, and experiencing more unexplained anomalies. A few "cleaning events"—winds that scoured the panels clean—came and went, increasing available solar power, but to little avail. Even as power output rose, it became increasingly clear that Spirit was stuck. Sandy soil had trapped the rover en route to the outcrops just ahead.

In early 2010, after months of trying every trick in their extensive playbook for rover driving, NASA declared Spirit a stationary research platform—in effect, a static lander. The probe could still perform valuable science where it sat, examining nearby soils and rocks, imaging terrain, and weather.

SPIRIT DEFEATED

On March 22, 2010, Spirit sent home its last message. It was nothing special—just routine data. The machine indicated that power levels, while minimal, were acceptable and that basic functions would continue as another winter edged closer. But after this message, no others were received. The engineers postulated that the rover had entered a deep hibernation mode, switching off to await better battery charging conditions upon emergence from the depths of winter.

By mid-2010, so-called "sweep and beep" operations were being used in attempts to wake the rover. These transmissions from the ground were timed at intervals during which Spirit was expected to be listening in, were it capable of doing so. Over thirteen hundred messages were sent to Spirit through May 2011, but were met with silence. What ill fortune befell Spirit was a mystery, but was most

The rover was heading uphill to an area known as Home Plate, but the dragging wheel was causing it to veer off course. To make matters worse, the slope was sandy, representing some of the most difficult terrain the team had yet to encounter. As driving conditions deteriorated, mission planners decided to reverse direction and drive downhill as a precautionary measure. There were attractive rock outcrops about 1000ft (305m) away, and they seemed like relatively safe options for the ongoing science program.

But it was not to be. As 2009 ground on, Spirit was encountering more and more difficulties. While it had driven 100ft (30m) at one point, and a few dozen more after that, energy was a problem again with dust on the solar panels. The onboard computers were

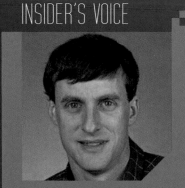
Steve Squyres was involved with the MER from the beginning. At the time, few could have guessed that it would become a decade-plus-long mission.

"The biggest discovery from Opportunity came in the first 60 sols, so we lucked out. We discovered a giant impact crater, that basically had all the things that we could have wanted, exposed right in the crater wall, within two months. Since then, we've taken advantage of the fact that Meridiani Planum is very smooth, very flat, and very capable for driving. We've been able to cover a lot of ground, and our strategy has been to go

from impact crater to impact crater. We're driving around on layered sedimentary rocks that were horizontally layered. They were structured, so you're basically seeing the same kinds of rocks over and over again. So what you need is some capability to get down to the rocks below the surface. We didn't bring a drill rig on this mission, but Mother Nature has made many craters for us on Mars. So we've gone to those craters, and finally to a big one, where we can explore what lies below the surface."

Opportunity's exploration of Endeavour Crater continues.

likely a combination of insufficient power, plunging temperatures, and wear and tear.

The mission was, however, in no way a failure. In fact, though the science returns from Spirit were considered to be less astounding than those from Opportunity, the story of the unlucky rover may have one final surprise in store. New studies of images taken by Spirit in 2008 have posed tantalizing questions and dangled delicious clues. In research released in late 2015, scientists took a second look at some oddly shaped structures spotted near its final resting place at Home Plate. These strangely shaped silica formations have been nicknamed "cauliflowers" for their unusual, lobate shapes. When first observed, these were thought to be typical of the types of formations that would be found near watery hot springs or geothermal vents, which are already pretty exciting things to see on Mars. But further examination of the eight-year-old photographs led the researchers to another possible conclusion.

These formations closely resemble similar structures found on Earth in the Atacama Desert of South America, the Taupo volcanic area in New Zealand, and Yellowstone National Park in the US. What these areas have in common is geothermal activity and related silicate structures that are thought to have resulted from microbial life. These structures strongly resemble those found in Spirit's pictures, and that is a nice final act from America's second longest serving planetary rover.

Spirit drove a total of 4.8 miles (7./km) over six years, many times the most optimistic projections of its primary ninety-day mission. It was the first sophisticated roving vehicle on Mars, it made

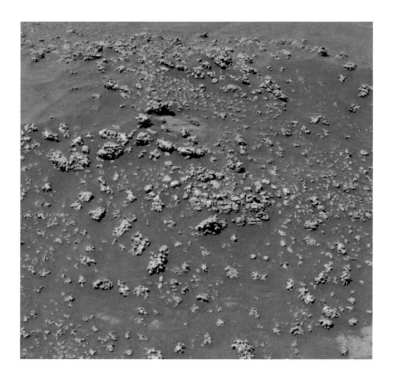

profound discoveries about the nature of the ancient environment, returned 124,000 images from the surface, and functioned twenty times longer than planned.

As the now inert Spirit collects drifts of sand on its windward side, Opportunity continues its journey across the Meridiani Planum, some 2100 miles (3380km) distant, unencumbered by the knowledge that its twin is no more.

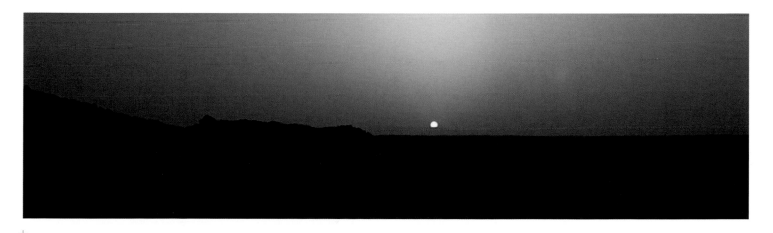

OPPOSITE: JPL engineers Joe Melko, to left, and Eric Aguilar perform traction tests with an MER twin test unit at JPL. Attempts to free Spirit from its sandtrap were ultimately unsuccessful.

ABOVE TOP: In 2008, Spirit returned this image of intricate formations of opaline near Home Plate which looked like cauliflower. Researchers looking at the images in 2015 noted a strong resemblance to microbial fossils found on Earth, raising the question of whether the Martian images might be indicative of the same thing.

ABOVE: In May, 2005, Spirit captured this view of a Martian sunset at 6.07pm on its 487th day, or sol, on Mars.

OPPORTUNITY'S GRAND TREK

It's tough to discuss the Opportunity rover without invoking the old television commercial about the Eveready Energizer Bunny. That advertising campaign may have worn thin after twenty-five years, but it is an apt metaphor for the incredible journey of Opportunity. The rover does, indeed, just keep going, and going. Opportunity is the longest-lived surface probe on any planet.

VICTORIA CRATER

Opportunity's arduous and thrilling trek has been five times that of Spirit and has been long, arduous, and thrilling. From mid-2006 until mid-2008, Opportunity explored the 0.5-mile/0.8km-wide Victoria Crater, a 230ft/70m-deep pit with an undulating wear pattern around its perimeter—it looked like an enormous protoplasm. The subsurface exposure of these worn fingers would provide a unique view into the planet's past, as well as gently sloping, safe access to the crater floor. Opportunity spent much of 2006 exploring the crater rim, peering cautiously into the depression beyond. Then, in September 2007, after a set of traction tests at the entry point, the rover headed down toward the sanddrift-covered bottom.

The drive into the crater had been preceded by yet another fortuitous wind-gust dusting of Opportunity's solar panels and a software upgrade, including improved navigation software. This had been tested with short autonomous drives along the rim, and would now contribute to a successful descent into the interior.

There were some slips and slides as the rover moved into the crater—this was tricky business. With Opportunity making the descent autonomously, at angles of up to thirty degrees of tilt, the robotic brain would stop, assess the topography, and make short lateral drives to test for better traction and less severe angles. After much trial-and-error, it made its way downward, stopping to look at and test a variety of targets en route.

For a few weeks, there had been concerns about Opportunity's robotic arm—it was using more electrical current than it should. The instrument-laden limb would occasionally reach a pre-set limit and "stall," but with small programming workarounds, was still able to accomplish most of its objectives. However, the engineers found that they did have to account for the tilt of the rover as it explored inside the crater. Even though gravity on Mars measures only 0.38

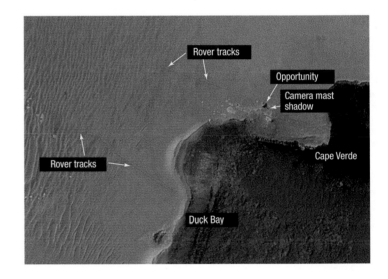

that of Earth, moving the arm "uphill" did require more torque, and thus more care.

As the rover continued its exploration of the interior of the crater, it was simultaneously conducting communication tests with the Mars-bound Phoenix Lander, which would be arriving in May 2008. The armada of active Mars machines was about to be upped by one more inhabitant.

Another problem suddenly arose—the RAT rock-grinding tool was experiencing difficulties. Engineers at JPL spent days with the twin Earth-bound test rover in their Mars simulator, and quickly

ABOVE: An image from the Mars Reconnaissance Orbiter shows Opportunity's travels around the rim of Victoria Crater.

OPPOSITE: Victoria Crater is a glorious sight, with scalloped sand dunes on its floor and highly eroded "fingers" reaching back from the rim, which provide views of the sedimentary layering on the surface.

devised alternative methods to approach the finicky task. Tiny things can rapidly become large problems if not carefully assessed when dealing with robots many millions of miles away. For example, if a grinding sequence is slowed or interrupted by a stall or negative feedback from the grinding motor, and will extend to the following day, the rover must be smart enough to back away from the rock during the Martian night. Winter was coming, and the overnight cold was severe enough to cause the metal in the robotic arm to shrink. The expansion caused by the machine's warming the following morning could cause enough movement to jam the brush into the rock face, bending the bristles. These are the kinds of problems that keep a bevy of engineers, programmers, and planners working, sometimes around the clock, to keep the rover safe and healthy. It is a job that requires deep passion for, and commitment to, the task at hand.

In April, Opportunity began a trek to the most magnificent site in Victoria, known as Cape Verde, a windswept bluff of exposed, strata-bearing bedrock. Along the way it would have to cross a few stretches of sand, which were slow going, as each patch had to be tested for drivability. If the wheels slipped more than expected, the machine could get stuck for good. Despite this careful navigation, Opportunity did experience a number of slips—just a few inches, but with experience gained from Spirit's adventures, enough to grab everyone's attention. With a successful crossing completed, imaging of Cape Verde began in late June. After a month of careful, high-resolution examination of the target, the rover headed out of Victoria Crater in August. The rover had only driven a short distance into the crater, but few expressed regret at leaving. The

drive had been harrowing and there was much more territory to be explored at Meridiani Planum.

ENDEAVOUR CRATER

As Opportunity prepared to head to Endeavour Crater, the next stop on the rover's tour of Mars, the engineering team took stock of the rover's health. They were still struggling with intermittent problems in one of the robotic arm's joints, and, like its twin, a front drive wheel was starting to act up. If either unit failed, the mission would be seriously compromised. Constant monitoring of both issues was performed as the rover began its long drive to Endeavour, over 7 miles (11km) distant as the Martian crow flies. That distance was approximately the same as the total mileage covered since Opportunity's arrival on Mars four years earlier, and the route the rover would be forced to follow, avoiding obstacles and seeking targets of opportunity, would be much longer. It was time to punch the accelerator.

As this new road trip was about to begin, the rover had to park itself for two weeks for what is known as "solar conjunction," a period during which Earth and Mars are on opposite sides of the sun and communication becomes unreliable. The safe plan was simply to park the rover until radio signals improved. Planners picked a benign but interesting rock outcrop that the robotic arm could study autonomously during the blackout and prepared to wait out the gap in communication.

When they were once again able to speak to the rover in December, the scientists discovered that the computer's memory

was overflowing with data, and needed to be purged. After this small delay, it was off to Endeavour again. The road ahead offered Opportunity a chance to make some record-breaking drives each day, and the rover was turned 180 degrees to drive in reverse and minimize stress on the ailing right front wheel, as Spirit had done.

Just as they began to make good progress, an unexpected Earth-bound problem arose. Fires in the Southern California hills that are the home of JPL caused day-long evacuations of the laboratory, leaving the rover to fend for itself until operators could return to work. Opportunity waited patiently.

After an epic two and a half year drive, Opportunity reached Endeavour in August 2011. Along the way it had zigged and zagged to investigate a number of interesting rocks and found a few more meteorites along the way, confirming much of what had been learned by Spirit's—and its own—investigations. Opportunity had driven almost 21 miles (34km) in seven years.

As August of 2012 approached, rover activites were minimized for a time while JPL mobilized for the arrival of the long-planned Mars Science Laboratory rover, Curiosity. That machine's complex but successful landing was accomplished on August 5, and NASA now had two rovers working on Mars. That was a new page in the history books, with Opportunity having entered its eighth year of surface operations a few months earlier. Now JPL's mission control would be running two separate machines on the surface of Mars, along with the Mars orbiters and all the other spacecraft under their control, from Earth-monitoring orbiters to the rim of the solar system, where the Voyagers were preparing to exit into

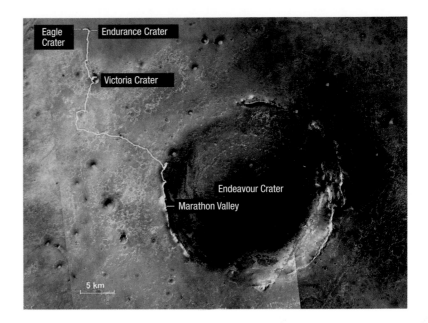

interstellar space. Never before had so much data been coming in simultaneously from so many missions.

As the rover drivers worked with the science teams to map out their objectives, one important target at Endeavour loomed

ABOVE: Opportunity has spent more time at Endeavour Crater than any other feature during its time on Mars, from 2011 to the present.

BELOW: A view of Victoria Crater from Opportunity. This panorama was completed in November 2006.

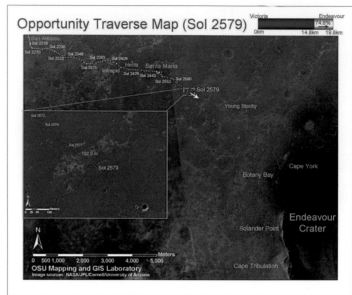

large. There was a grouping of phyllosilicate rocks that had been detected from orbit. These are minerals like talc and clays that have interacted with water in their formation. While both rovers had investigated plenty of rocks associated with once wet conditions, each new finding helped to draw the new map of Mars that included an ancient, watery environment. Mapping out their positioning in the layering of Martian rocks, verifiable only at close range by the rovers, also helped the geologists to develop a timeline for Mars' ancient environment. With no bodies of standing water, nor any streams or creeks, on present-day Mars, it was important to build a history of when atmospheric conditions—temperature and density—might have supported standing, liquid water. This would establish a likely period for the possible existence of life on the planet—if there had ever been any.

From the rim, Endeavour was an imposing sight. At 14 miles (23km) in diameter, it was by far the largest feature visited by the MER rovers. It has also been one of the most interesting areas to be found on Mars, which accounts for the rover having spent five years exploring the rim. Mission scientists referred to Endeavour as "Landing area two" and "A brand new mission." Not only did the surrounding region include the orbitally observed phyllosilicates, but other watery rocks as well. And the samples examined in this area were older—much older—than those seen

before. They also showed that Mars had been even more watery than the previous samples had shown. Endeavour Crater is over 3.5 billion years old, and the rocks there indicate not just a wetter environment, but one soaked in more benign water—not so acidic or alkaline as what was seen earlier in the mission.

The overwintering of 2013–14 saw the rover exploring a bluff named Solander Point, where it eventually parked, angled toward the sun to collect any available light it could find. Since then, the rover has imaged Martian moons, comets, and of course, many dozens of rocks and patches of soil. In all, almost 300,000 images have been sent home to Earth, along with immeasurable amounts of scientific data.

Opportunity is still exploring the rim of Endeavour Crater today, traversing an area named Marathon Valley. Like a retired prizefighter, the rover shows signs of age and scars from its adventures. The robotic arm no longer moves from side to side, so anything the scientists want to investigate must have the arm maneuvered to it via careful driving. The Mossbauer spectrometer gave out some time ago, and one of the front wheels is slightly rotated inward—Opportunity is pigeon-toed. Finally, since 2013, Opportunity's electronic memory has become increasingly faulty and problematical. In sum, it's just plain harder for the mission planners to use the rover effectively… but they do.

After twelve years, Opportunity just keeps going.

ABOVE LEFT: Vandi Tompkins poses with an MER rover at JPL's Mars simulation yard. Tompkins is a robotics engineer and rover driver at JPL.

ABOVE RIGHT: This map shows the travels of Opportunity up to 2010. The rover continues to explore the area around Endeavour Crater to this day.

TOP: In October, 2010, Opportunity sent back this image of the rim of Endeavour Crater. This image has been color processed to reveal increased detail. The mountains in the distance are the rim of the crater 19 miles (30km) distant.

ABOVE: This 2004 image combines a view of Meridiani Planum with a computer-generated image of Opportunity. Called Virtual Presence in Space, the JPL-invented technique gives a sense of scale to surface images.

HD IN SPACE: MARS RECONNAISSANCE ORBITER

The Mars Global Surveyor orbiter had been a fantastic success, sending back the best images yet of the red planet in unsurpassed detail. The Mars Odyssey mission had then gone on to provide vastly improved remote sensing and visual data that changed how we looked at Martial topography and allowed mission planners to determine landing sites for both MER rovers. However, both machines were showing their age by 2005, so it was welcome news when the Mars Reconnaissance Orbiter (MRO) launched successfully in August of that year.

AS A PART OF THE OVERALL MARS ORBITER PROGRAM, MRO was a logical, yet daring, advance over its predecessors. It was the logical successor to the previous orbiters—similar instrumentation, improved cameras, and some unique new experiments. But the most immediately noticeable change from its Martian siblings was the enormous 10-ft (3-m) radio dish. MRO would be sending more data than any previous Mars orbiter—high-resolution images, scientific telemetry, and continuing relays from the MER rovers, the Mars Phoenix polar lander slated for 2008, and the future Mars Science Laboratory Rover, which would land in 2012. MGS (which would fail just seven months after MRO's arrival and only two months after it completed aerobraking) and MO were already operating at their limits, to accommodate the combined data loads of their own instrumentation and the MER rovers on the ground. The addition of Phoenix and MSL in a few years would swamp the orbiting network, even with ESA's Mars Express helping out. The large antenna and its huge transmission capability were critical to ongoing operations.

NEW TECHNOLOGY

However, MRO was no mere relay station. The orbiter had a wonderfully advanced set of new exploratory tools as befitted its comparatively generous budget, which was in the region of $720 million. Of course, it had a spectrometer to complete mineralogy studies of the Martian surface—that was a given. It was called the Compact Reconnaissance Imaging Spectrometer for Mars (CRISM), and would provide the most detailed readings of the surface makeup yet.

A second spectrometer, the Mars Climate Sounder (MCS) was a spectrometer, and was specifically designed to probe not the surface but instead the layers of the atmosphere, in 3-mile (5-km) sections. MRO also carried a powerful radar unit named the Shallow Sub-surface Radar (SHARAD), to probe over 0.5 mile (800m) below the surface; deep ice deposits at the polar regions were of particular interest. But, as wonderful as these instruments were, the cameras were the real stars of MRO.

Three separate sets of cameras rode along on the orbiter. MARCI (Mars Color Imager) was a wide-angle camera that provided regular weather reports from visual data—the entire planet was imaged daily. HiRISE, the High Resolution Imaging Science Experiment, was the pinnacle of optical design—a true telescopic camera, the largest ever flown to Mars, with a 20in/51cm-diameter mirror. It was five times more powerful than the camera on MGS, and would be capable of discerning objects as small as 3ft (1m) across. As a general comparison, where previous orbiters could pick out something the size of a city bus, HiRISE could image objects smaller than a go-cart. The data-loads from this instrument alone, with its

BELOW: Mission patch artwork for the Mars Reconnaissance Orbiter.

MARS RECONNAISSANCE ORBITER

MISSION TYPE: Mars orbiter
LAUNCH DATE: August 12, 2005
LAUNCH VEHICLE: Atlas V

ARRIVAL DATE: March 10, 2006
MISSION TERMINATION: Continuing

MISSION DURATION: 11+ years
SPACECRAFT MASS: 2273lb (1031kg)

ultra-high definition images, merited the large radio dish on MRO.

The third camera was called the Context Camera or CTX, and was a unique addition. With a wide-angle design, and a resolution somewhere between that of the MARCI and the HiRISE, the CTX provided a steady stream of images that could be matched to both the images taken by HiRISE and readings taken by CRISM, to provide a visual context for those instruments. The value of simultaneously taking close-up and wide-angle images of the same area was immense; knowing what the wider view of something interesting seen in a close-up from HiRISE, or a peculiar reading from CRISM, proved to be critical to many new discoveries.

MRO reached Mars in 2006 and slowed itself into a lopsided orbit using its rocket motors. The spacecraft carried a load of just over 300lb (136kg) of propellant—enough for the orbital insertion around Mars and at least a decade of maneuvering. After the obligatory six months of aerobraking, the 2273-lb (1,031-kg) science platform began its primary operations. MRO orbits Mars in just under two hours at an average altitude of about 175 miles (282km) above the planet. The mission was designed, in general terms, as an extension of the MER rovers and Mars Odyssey orbiter: a continued search for ancient watery regions and subsurface ice. Of primary interest were water-created minerals, ancient shorelines, lakebeds, and sedimentary deposits. These science targets were derived from examination of previous orbital missions—in particular, the well-studied imagery from the Vikings and MGS, and, to a lesser extent, rover data. Yet, with MRO's triple cameras and data capability, the number of regions investigated rapidly reached over ten times that of all previous surface surveys.

Besides the surface science mission, future landing zones were mapped with care. Finally, mission planners could actually see medium-sized boulders instead of having to intuit them from surrounding terrain, and this ability made for much better calculations of rocks in the 1-ft (30-cm) range—not big enough to kill a large lander, but big enough to jeopardize its function. MRO represented a huge step forward in cherry-picking the low-hanging fruit of regions rich with geological treasures for closer investigation.

As plans for the 2008 landing of the Mars Phoenix Lander were being finalized, the first-choice landing site was imaged by HiRISE and found to be rife with boulders. The high-resolution cameras also allowed the orbiter to do some sleuthing: the remains of the failed Beagle 2 lander from Mars Express, which crashed in 2003, as well as the carcass of the Mars Polar Lander from 1999, were both spotted. Both MER rovers have been photographed numerous times, and MRO assisted in selecting safe drives for Opportunity and Spirit at each major objective.

GROUNDBREAKING DISCOVERIES

In late 2006, the MCS atmospheric spectrometer began to malfunction; workarounds were explored and used with some success. The orbital environment is tough on spacecraft and electronics, and radiation in particular takes a toll. Some pixels have been lost in the CCDs in the HiRISE camera, but such failures are expected over time, and if they are minor enough to allow measurements to continue, the science team is grateful. There was a heart-stopping moment in 2009, when the onboard computer began a series of reboots. JPL engineers put the spacecraft into safe-mode, effectively suspending the mission for four months. Over a hundred suspected causes were evaluated, and while the cause was never conclusively determined, it may have been due to the radiation in the Martian environment—as with previous spacecraft, binary 1s and 0s can be reset by stray high-energy particles, corrupting software. By 2010 things were back to normal, but with a watchful eye on the computer.

The discoveries credited to MRO would fill an entire book, but a short list includes:

- Multiple observations of new craters made by recent meteor strikes, some of which showed that water ice was thrown up and around the crater. This rapidly disappears, but also allows the scientists to add these regions to places with known water deposits.
- Deposits of chloride minerals, or salts, have been spotted in a number of areas. These are thought to have been left behind by bodies of mineral-rich water that have evaporated in the distant past, much like crystal-laden shorelines around salty lakes or ponds on Earth.
- Other types of minerals indicative of past water, including clays, have been found over large areas of Mars by the CRISM instrument. By now this was not a shock to anyone—Mars was understood as having had plenty of water in the past—but the broad distribution and amount of it was surprising.

There have even been "live" events captured by MRO's cameras.

- A number of avalanches—complete with huge, red dust plumes—have been imaged, and after meticulous planning, the

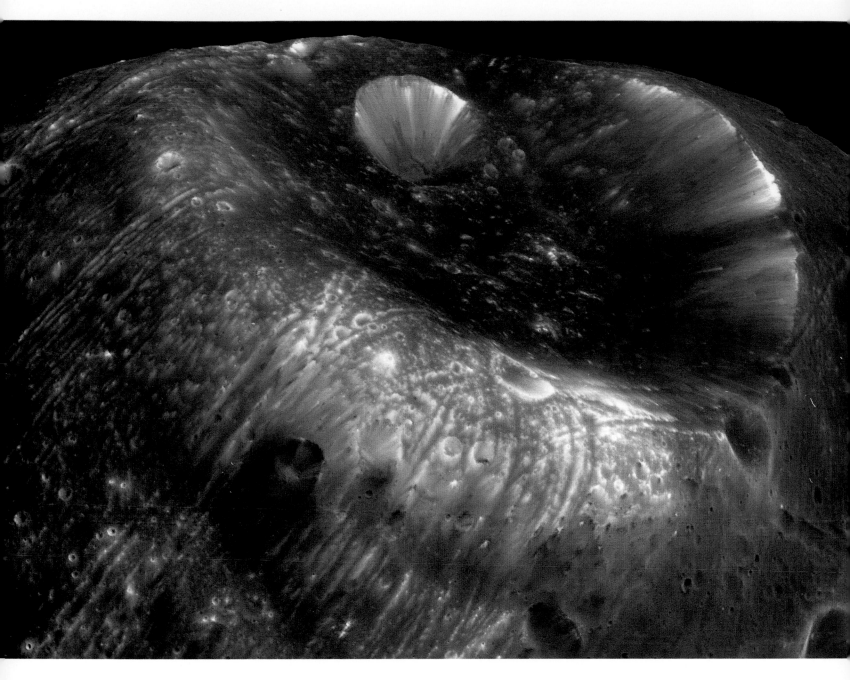

landings of both Mars Phoenix and the Curiosity Rover were captured as they parachuted to the planet's surface.

• The HiRISE camera has taken stunning pictures of both Martian moons, Phobos and Deimos, in unprecedented resolutions. These have assisted scientists studying those bodies for both geological curiosity and for future mission planning—it may be that we will land humans on the Martian moons well before Mars itself.

• Minute weather-attributable changes have been imaged on the surface, including traces of the paths of dust devils crossing the surface. These wind-driven vortexes can be upwards of 12 miles (19km) high, and sometimes leave fantastic patterns on the sand. Their frequency, size, and direction can tell researchers much about atmospheric structure and wind patterns.

• In perhaps its most remarkable finding, MRO imaged dark streaks on some Martian hillsides in 2011. These were studied for years, and in 2015 NASA announced that liquid water had been found on Mars. The briny fluid seems to seep out from bluffs and canyon walls in certain weather conditions, barely wetting the nearby soil, but when a planet is as parched as Mars is, any liquid water is a welcome observation.

OPPOSITE: The elusive Martian moon Phobos shows off its largest crater, called Stickney. The image was taken when MOR was about 4,200 miles (6,800 km) away.

RIGHT: The Earth and moon as seen from Mars in 2007. The spacecraft was already in orbit around Mars, and was 88 million miles (142 million km) from Earth.

BOTTOM LEFT: Tracks from multiple dust devils can be seen crossing Martian dunes in this 2009 image. These fresh tracks will be gone before long.

BELOW LEFT: HiRISE captured the parachute of the Curiosity rover's Seven Minutes of Terror on August 5, 2012. MRO captured this one-chance shot.

BELOW RIGHT: In 2014 this image, and others like it, generated excitement when scientists realized that they were looking at recent flows of water seeping out of canyon walls at Palikir Crater. Since then others have been observed, and the announcement of the discovery was made with great fanfare in 2015.

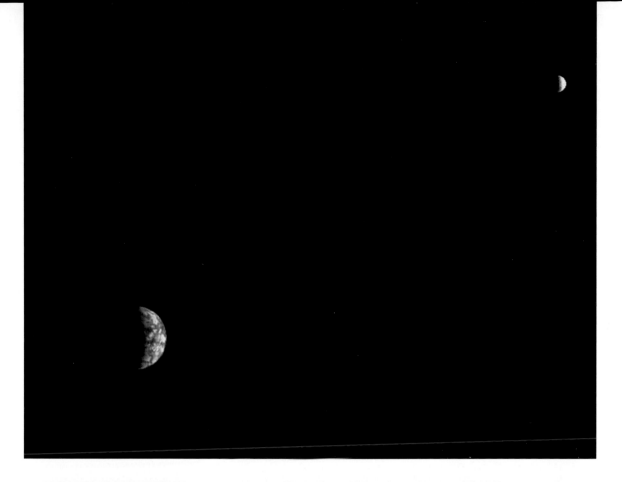

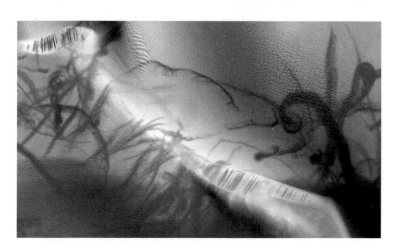

Richard Zurek
Chief Scientist, Mars Reconnaissance Orbiter

"When you look at mission like MRO, it's on the order of the great discoveries of Mariner 9 and the Viking orbiters and landers. Discoveries begin to build up on you incrementally, but they are profound. One of the things that really impressed is that the ground is so quizzically patterned. We saw polygons and fractures—it looks like a surface that has had many places that were once wet then dried out. It's like looking at mudflats on Earth, only the scales are bigger than that. I think that tells us a few things. First, there have been drying out episodes; second, that there's still ice within the surface of the planet; and third are the composition measurements, that have indicated that you have areas where certain minerals are present. When you put that whole picture together, what you see is an ancient surface that's been covered up, and where there was a lot of water interaction. This altered the composition of the surface … which is interesting, the fact there are different minerals indicating different kinds of watery environments, some of them more acidic than others. To me, this once again increases the potential that Mars might have developed life sometime in its history, and that's pretty exciting."

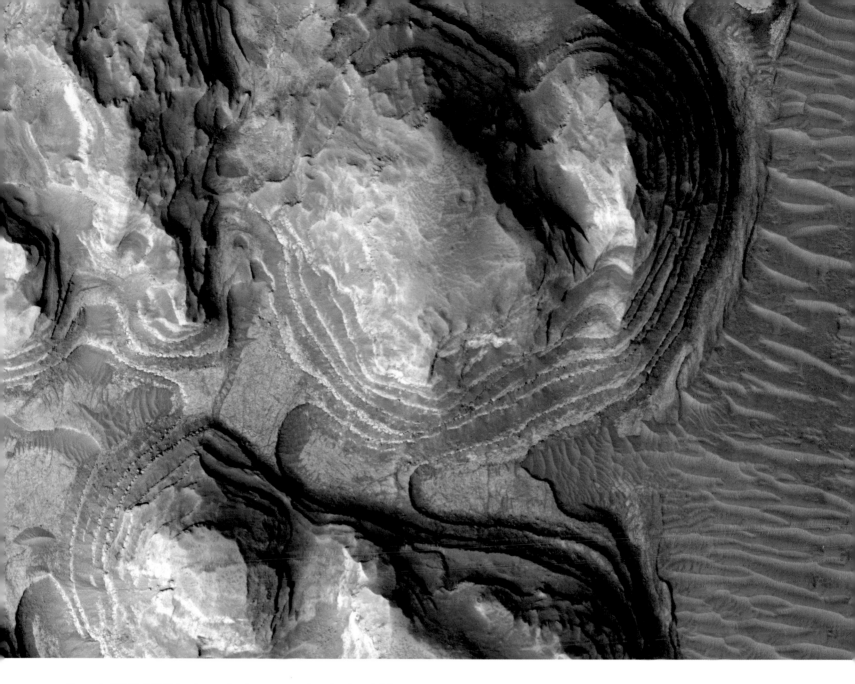

By early 2015, MRO had sent back over two hundred and fifty terabits of data, much more than all previous planetary missions combined and equivalent to fourteen billion printed pages of data. As the orbiter was investigating the surface, it would also record uplinks from the landers and rovers on the surface and store them for later playback to Earth, when conditions were most favorable.

In a decade of operation, and in its fourth mission extension (as we have come to expect from JPL's Mars spacecraft), MRO has rewritten parts of the Mars history book and filled out others. Visual and spectrographic measurements have revealed three major periods of Martian history. The oldest, seen on the densely cratered highlands, demonstrates extensive evidence of vast watery environments, later validated in part by the rover missions. A later era shows water active in the atmosphere, cycling between polar ice and lower-latitude ices and snowfall. Then comes the modern period, characterized by a dry atmosphere, copious amounts of carbon dioxide, and small water cycles. MRO's mission continues, as do the discoveries and vast quantities of data for later study. It will eventually be moved to a higher, more stable orbit, from which it may operate well into the 2020s.

OPPOSITE: Taken in late spring, this shot shows a covering of fine-grained carbon dioxide frost (with a dusting of water ice) receding from a nearby crater.

ABOVE: An example of "cyclic bedding" in the Arabia Terra region. The steep sides of the buttes, and their stepped nature, may be due to alternating softer and harder layers of rock eroding at different rates.

EMPIRE OF ICE: MARS PHOENIX LANDER

The Mars Phoenix Lander mission was one of many firsts: first lander to reach a polar region of Mars, first mission controlled out of a university, and, perhaps most remarkably, first probe to make direct contact with water on Mars. Phoenix was a part of the Mars Scout program, another of NASA's lower-cost exploration initiatives. This was no "faster-better-cheaper" fantasy though; the budget cap was $485 million and could be done with some clever planning.

SINCE THE PHOENIX LANDER was basically a reboot of the 1999 Mars Polar Lander attempt, it made sense to fly it under the Scout program banner. As it turned out, Phoenix would come in at well below the budget cap. This was in no small part due to the unusual control and operations structure applied to the mission.

AN UNUSUAL MISSION

Some Mars missions seem to be driven by one or two strong personalities… for example, with Mariner 4 it was Robert Leighton (see page 19), for the MER rovers it had been Steve Squyres (see page 100). Phoenix was proposed—and won—by Peter Smith of the University of Arizona. His association with NASA's Venus missions in the late 1970s, and his resumé included building the primary cameras for the Mars Pathfinder and Mars Polar Lander missions, as well as the HiRISE camera for MRO. When he proposed the Mars Phoenix Lander, Smith was a natural to win the competition for NASA's dollars. It was a tremendous opportunity to be the Principal Investigator for a lander, and a way to recover the science lost when the MPL mission had crashed a few years earlier.

The chosen landing zone was in an area named Green Valley, which contained the largest known water ice deposits outside the polar icecaps themselves. All this was in a region called Vastitas Borealis—translating as the much less scenic-sounding Northern Waste. This is a part of the northern plains or lowlands, the great expanse of smoother terrain that makes up the "top" half of Mars. Phoenix set down near the polar border of what is now suspected to be an ancient and large northern ocean. The site had been chosen after data from Mars Odyssey showed huge deposits of water ice extending down beyond the borders of the pole, as inferred from gamma-ray spectrometer readings of hydrogen beneath the soil.

The lander was contracted to Lockheed Martin, the company responsible for so many Mars spacecraft. However, Smith kept the instrument design and construction close at hand to control costs, building the camera at the University of Arizona and contracting out other instruments to universities around the world. However, perhaps the most unusual aspect of the mission was the control center: it was housed right on the UA campus. In a small building appropriated for the task, worktables were set up with laptops and endless cable-runs, to provide the needed control center to run the mission once it landed. JPL would fly and land Phoenix, but it would be run from the university, largely by graduate students—it was the first time a public university had assumed such control over a space mission. It was a brilliant and effective way to keep costs under control and provide maximum return on investment.

One other mission parameter assured controlled costs. While the lander had the usual ninety-day life expectancy of many modern Mars missions, it was unlikely to exceed this amount by years (with the associated costs) like so many of JPL's missions have. Phoenix would be landing near the north pole of Mars, and the crushing cold there would almost certainly end the mission as local winter set in. In the end, it did exceed the ninety-day requirement, but only by a little over two months. This was expected, and every day over the first three months was a gift.

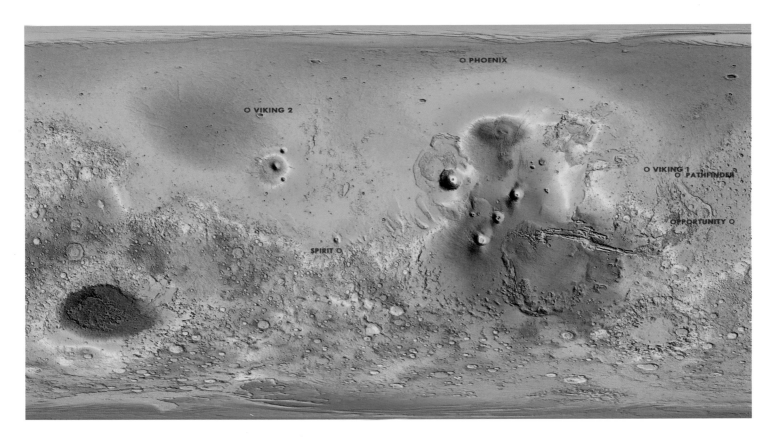

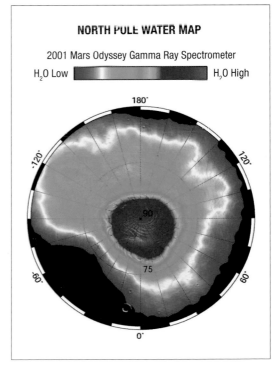

ABOVE TOP: Mars Phoenix landed in a region called Green Valley near the north pole, at about 60 degrees latitude.

ABOVE LEFT: Technicians at Lockheed Martin, the builder of the Phoenix lander, work on the spacecraft nestled inside its aeroshell.

ABOVE RIGHT: This image from the Mars Odyssey orbiter shows why scientists were so interested in a mission to the polar region—water ice, and lots of it, was apparent.

The design of the lander was based on previous designs—the failed Mars Polar Lander and the never-flown Mars Surveyor 2001 plans. A flat platform with three landing legs was flanked by two circular, folding solar panels, and a single robotic arm would provide samples for the instrumentation within. The arm was capable of extending about 7ft (2.1m), and was expected to be able to dig about 16in (40cm) into the soil. A rotating grinder (somewhat akin to MER's RAT tool but tougher) would dig into the ice to provide access to what lay below.

The instrumentation included the Surface Stereo Imager, a dual-camera system not unlike that flown on Pathfinder and the MER rovers. Two microscopes were onboard, one optical microscope, and an Atomic Force Microscope, with the ability to analyze a single grain of dust at a time. A final camera, the Mars Decent Imager (MARDI), would record the landing as seen from the bottom of the spacecraft.

The Thermal and Evolved Gas Analyzer (TEGA) instrument contained a high-temperature furnace with a mass spectrometer, capable of analyzing the composition of soil samples when slowly heated up to 1800°F (982°C). A "wet chemistry lab"—the Microscopy, Electrochemistry, and Conductivity Analyzer (MECA)—was a surplus design from the Mars Surveyor 2001 mission. MECA had four chambers that would mix water with soil samples and measure the biological compatibility of the Martian dirt.

A probe mounted on the end of the arm, called the Thermal and Electrical Conductivity Probe (TECP), was capable of measuring the temperature and humidity of Martian soil, as well as the water vapor pressure of the atmosphere (when not embedded in the ground). Finally, a meteorology station (MET) provided daily weather reports of the environment, including the levels of dust contained in the cold air.

PHOENIX RISES

Phoenix was small and lightweight as such things go, tipping the scales at only 779lb (353kg). It launched in August 2007 and successfully landed on Mars at the end of May 2008. The available orbital assets—that is, other Mars spacecraft—which included Mars Odyssey, the Mars Reconnaissance Orbiter, and Europe's Mars Express, were all positioned to track the landing and relay data as needed. This not only allowed for improved communications, but also provided an exact location of the final landing site. The mission planners for MRO in particular spent a lot of time preparing to capture the descent of Phoenix—no small feat, given the long delay of radio signals between Earth and Mars—and successfully snapped a picture of the lander as its parachute delivered it to the surface. This feat was akin to snapping a picture of a speeding car on

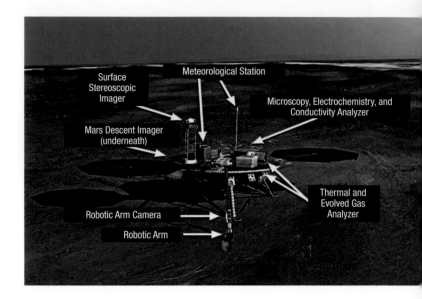

a freeway, merely by knowing where it departed from, where it was supposed to go, and the general speed at which it was traveling. It was a sensational achievement.

There were a few sweaty-palm moments. This was a difficult trajectory—all previous landers had set-down much closer to the Martian equator. Since Phoenix was landing near the pole, the entry to the atmosphere was a bit trickier. Also, this was the first landing to use rockets to slow the spacecraft all the way to landing since Viking (which had a vastly larger budget in adjusted dollars). Pathfinder and the MERs had used airbags to bounce to a stop. The only other attempt at a rocket-powered soft landing had been the Mars Polar Lander, and everyone was well aware of how that had turned out (see page 76). As the budget-priced lander hurtled through the thin Martian air, the pressure was on....

ABOVE TOP: The compact and lightweight lander design of Mars Phoenix was a descendant from the Mars Polar Lander and would be recycled for the InSIGHT lander, scheduled for a 2018 launch.

ABOVE: The landing zone was surrounded by polygons, apparently caused by thermal shifts that cause the ground—similar to Earthly permafrost—to expand and contract.

MARS PHOENIX

MISSION TYPE: Mars lander

LAUNCH DATE: August 4, 2007

LAUNCH VEHICLE: Delta II

ARRIVAL DATE: May 25, 2008

MISSION TERMINATION: November 2, 2008

MISSION DURATION:

SPACECRAFT MASS: 770lb (349kg)

When the time came for the parachute to deploy, to slow the spacecraft to reasonable speeds, that time marker came and went—with no indication of parachute release. The signal came in seven seconds late—not a lot of time, but an eternity to someone sitting at the console tens of millions of miles distant. The lateness of this event also caused Phoenix to overfly the center of its landing ellipse, meaning the spacecraft arrived right at the edge of the zone. It was perfectly acceptable, but not what they were used to at JPL. The Galactic Ghoul (see page 27) had tried to snag this one, but Phoenix narrowly escaped his grasp. But most importantly, it had landed safely and relayed this event to cheering and applause on Earth.

Within short order the solar panels deployed properly, the meteorology instruments began their work, and the cameras imaged the surrounding terrain. It was a landscape unlike any previously seen: the rocks were smaller, the surface almost flat, and perhaps most remarkably, the terrain was crisscrossed by patterned lines—the entire area was covered with flat polygons. These angular shapes were between 6–14ft (1.8–4.2m) in diameter and bordered by small depressions caused by seasonal temperature fluctuations. The soil would expand and contract laterally, leaving the oddly shaped gaps not unlike those found on Earth's permafrost regions. They were fresh too—there was no weathering on the sharp edges.

There was a day-long delay using the robotic arm, when MRO's radio refused to relay the commands down to the lander. But Phoenix had back-up programming that it would follow if it did not receive commands from Earth, but before these could be invoked the issues with MRO were resolved and the instructions re-sent. The next day the arm went to work.

Almost immediately, electrifying images arrived, taken underneath the lander by a camera on the end of the arm. Broad white patches looking like polygonal white road signs were seen just beneath the landing rockets. And they were not small, measuring roughly 1ft

INSIDER'S VOICE

Peter Smith
Principal Investigator, Mars Phoenix

"I had spent a lot of time thinking about looking for life on Mars: where would I look, under what rock, how would we get there, and what would we do. At about the same time, one of the professors in our department had published a paper about finding ice under the northern plains of Mars with Mars Odyssey orbiter. He was able to probe about a meter deep under the surface using gamma rays and neutrons, and see that there was a solid ice layer right under that surface.

"If we could go there and understand the history of that ice, and the minerals and the chemicals that are in association with it, that would be an incredible mission. Could this region be a place like here on the Earth, where you get into these Earth analogs, like Antarctica and in the permafrost regions? The permafrost is the deep freezer of the Earth, it's where things are preserved. In the ice of the northern or southern permafrost regions you can find evidence of past life going back millions of years.

"This led me to think that perhaps this happened on Mars, and that might be a good place to look, so we wrote our mission goals around the permafrost of the northern plains of Mars."

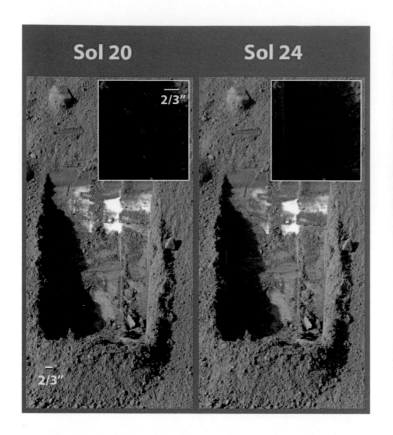

Sol 20 Sol 24

2/3"

2/3"

(30cm) across. The conclusion was that these were tabular ice formations, swept clean of loose soil and dust when the landing rockets disturbed the surface. There were likely hundreds more surrounding the lander.

After a few days of testing, which included depressing the soil and other tests for hardness, the arm started digging trenches in the icy ground. One of the early trenches dug by the scoop on the arm was named Dodo, and it was soon joined by a parallel trench called Goldilocks. Within minutes of seeing the pictures of the trenches, the scientists realized that they had struck clumps of something hard and white, after they had scraped off the loose soil. Most of us would excitedly yell: "Hey, it's ice!", but that is not how Mars missions are conducted. While excited, the team claimed no certainty of what they were looking at, just that it might be water ice. They kept an eye on the furrows. After a few days, the white clumps had vanished—it had been ice after all, and had evaporated, or more accurately "sublimed" away, due to the exceedingly low atmospheric pressure. The relatively slow rate at which it disappeared indicated that it was water ice; dry ice (composed of CO_2) would have vanished much more quickly.

Soil samples were delivered to the onboard chemistry laboratory for analysis. This was literally a sticky situation—the soil was clumpy and would not sift through the screens covering the collection funnels. It took a number of tries, but the mission scientists eventually figured out a way to vibrate the scoop and drizzle the dirt past the screen and into the test chamber.

The analysis of nearby soil showed the local samples to be moderately alkaline, and confirmed the presence of perchlorate in the dirt, as had been speculated since the Viking missions. Then, a problem—there was a short circuit in the TEGA. This was not immediately crippling to the instrument, but the engineers determined that it could not occur again when the doors covering the TEGA were opened.

So NASA decided to take an investigative shortcut. They would eventually make more in-depth sample analyses, but if there were going to be issues with TEGA, they had better perform testing for water right now. This was an easy decision to make—the mission was probably half over, by best estimates of the coming winter freeze. But harvesting the ice sample for analysis proved more challenging. The sample collection was harder than envisioned—the targeted ice was as hard as rock. However, within a couple of weeks the scientists had their ice sample and, after a bit of drama at the funnel again, got it into the TEGA.

Fairly quickly, the results came in—there was water in the soil, and just over two months into the primary mission, NASA made the announcement: Phoenix had made the first verified observation of water—not in the past, but now—on Mars!

Besides the ice discoveries, calcium carbonate and other chemicals indicative of earlier flowing water were found in soil samples, increasing the range of verified ancient liquid water bodies into much higher latitudes.

As the mission proceeded into winter, the environment reacted mostly as expected, with rapidly declining temperatures. However, there was a wonderful surprise when, one day in late September, water-ice snow was detected in the sky above the spacecraft. This was a first, and a sure sign that winter was fast approaching. Wind speeds averaged about 22mph (35km/h) with a peak of near 40mph (64km/h) and temperatures, which had reached a high of about 0°F (-18°C), were plunging toward a low of -142°F (-97°C).

GOODBYE PHOENIX

The remaining research projects were hurried as much as was possible, but the cold was draining the batteries faster than the solar panels could recharge them with the weak winter sunshine. On the 155th day of the mission, Phoenix fell silent. Attempts were made for weeks to regain contact, but most of the engineers were not optimistic. The lander had never been designed to last through the winter, especially up near the pole. Eventually, the following spring, the MRO orbiter snapped an image of Phoenix and it was apparent that one of the solar panels had collapsed, probably under the weight of accumulated ice. There would be no regaining contact with the lander.

Despite the short surface operations, Phoenix had been a grand success. It was the first lander to set down near a polar ice cap and the first to land under power since Viking. This was the first time that water had been detected on Mars, albeit in a frozen state, and the first mission to be run by a public university. Like Pathfinder before it, Phoenix had accomplished great things, and at a budget price.

As the science teams continued examining the data collected during the mission, sheets of ice began forming on the dead lander's solar panels. But back at JPL, attention had shifted to something new, and very special. The Mars Science Laboratory was nearing completion, and it would become one for the record books.

OPPOSITE TOP LEFT: After a trenching revealed a few patches of what looked like water ice (as opposed to CO_2), scientists saw it slowly disappear at a rate that made it clear that the substance was water.

OPPOSITE TOP RIGHT: This image was taken at 6am local time 79 days after landing. Though the mission was only half over, overnight frost was forming on the ground.

OPPOSITE BOTTOM: Mars Phoenix prepares to take a sample with its robotic arm. The solar power panel, which unfolds like a circular fan, is lower left.

SEVEN MINUTES OF TERROR: MARS SCIENCE LABORATORY

Not since the Viking missions of the 1970s had anything on this scale been planned for Mars. Everything about the Mars Science Laboratory—ultimately named Curiosity—was bigger and more complex. It was the most daring mission to the red planet to date, incorporating everything that had been learned from previous rovers and orbiters to optimize the investigative hardware. MSL contained what would have been, just a couple of decades before, two roomfuls of laboratory equipment in one economy car-sized rover.

SOME STATISTICS TELL THE STORY:

WEIGHT: 2000lb/907kg (the MER rovers Spirit and Opportunity were 400lb/181kg each)

INSTRUMENT PACKAGE WEIGHT: 276lb/125kg (MER's weighed 15lb/7kg per rover)

LENGTH: 10ft/3m (the MERs were 5ft/1.5m long)

CLIMBING ABILITY: MSL can crest obstacles 3ft/90cm high (MER could overcome 18in/45cm obstacles)

POWER SUPPLY: Nuclear, plutonium 238, continuous (MER used solar panels that worked only when the sun shone and it was not too dusty)

HEAT SHIELDING: MSL's heat shield is almost 15ft (4.5m) in diameter (larger than the Apollo Command Module's shield, which was just under 13ft/3.9m)

COST: MSL cost $2.5 billion by the end of its primary mission (the MERs were about $820 million)

HI-TECH TO MARS

But there was more to this new spacecraft than just larger mass and bigger numbers. While Curiosity looks generally like a scaled-up MER rover, it is a vastly more complex machine, virtually a chemical and geological laboratory on wheels. The instruments on Curiosity would be capable of ingesting samples, much as Viking and the Mars Phoenix Lander had done, and then running them through a battery of analyses that would determine their chemical make-up, right down to the isotopic numbers of each element identified. The primary goal of this was to seek organic compounds in the rocks and soil of Mars, which can be an indicator of past or present life, or at least an environment conducive to life. It was not equipped to seek out and identify life itself, however; merely the organic compounds that can indicate that it may have existed. Past and present conditions were the prime directive of the mission, and Curiosity was brilliantly appointed to ascertain the existence of such past and present conditions.

The rover sported some familiar, though vastly improved, instrumentation. A forward-mounted mast carried MastCam, a set of newly developed HD cameras capable of both still frames and video in 3D. There are two sets of optics, one wide-angle and one telescopic. Also mounted on the mast is the camera for the Chemistry and Camera Complex (ChemCam), a telephoto imaging spectrometer that would employ a powerful laser to briefly heat patches of rock beyond the rover's reach, up to about 20ft (6m). The camera would read the brief flash from the thin layer of vaporized rock, and extract information about its make-up from the spectra observed. Finally, two smaller black-and-white navigation cameras called NavCams are also mounted on the mast to facilitate driving on Mars, as is a meteorology station named the Rover Environmental Monitoring System (REMS).

A robotic arm extends from the front of the rover, similar to the MER design. The end of the arm hosts a cluster of investigative tools. These include:

• The Mars Hand Lens Imager (MAHLI), which provides micron-level photos, allowing geologists to see the granular make-up of rocks and soil.

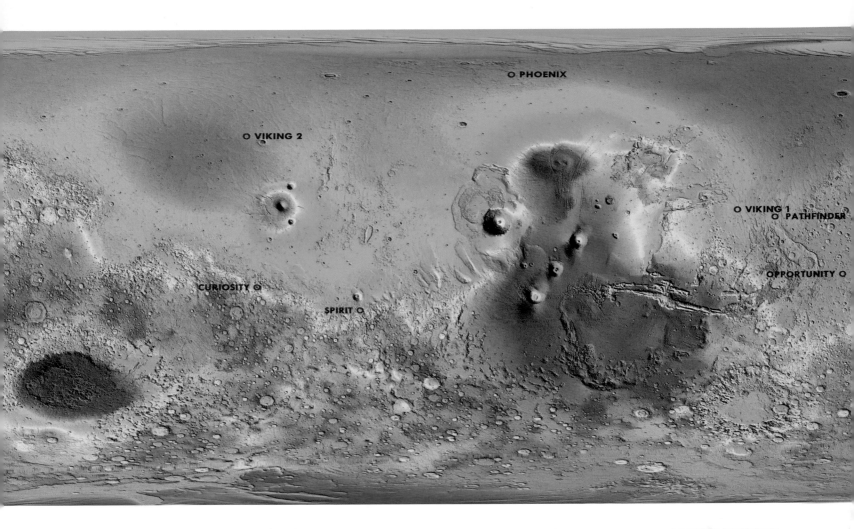

VIKING 2 O

O PHOENIX

CURIOSITY O

SPIRIT O

O VIKING 1
O PATHFINDER

OPPORTUNITY O

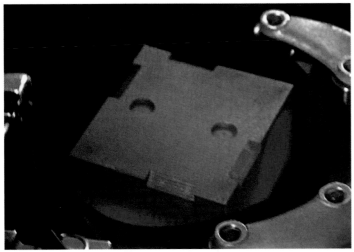

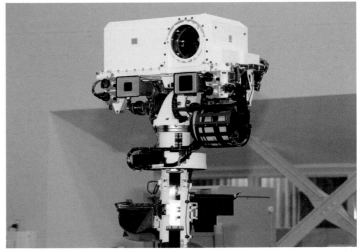

ABOVE TOP: Curiosity landed by plummeting into a crater near the boundary between the Martian highlands and the basaltic plains.

ABOVE LEFT: Red-hot plutonium 238 is the fuel powering Curiosity. It has a half-life of 88 years, but these power units often begin to falter after 14 years due to other issues.

ABOVE RIGHT: The camera mast for Curiosity contains ChemCam, the laser-powered spectrometer, top, and MastCam, comprised by the smaller lens sets just below the white box.

- An Alpha-Proton X-ray Spectrometer (APXS), an improved version of the APXS units that all previous rovers carried.
- A rock-cleaning brush, a new and unique percussion drill, and a scoop/sieving mechanism for delivery of soil and pulverized rock samples to the onboard laboratory.

On the body of the rover are found the Dynamic Albeldo of Neutrons (DAN) experiment, which measures water near the surface below the rover, and the Radiation Assessment Detector (RAD), designed to provide continuous readings of the ambient radiation in the environment. But inside the body of the rover is where the real magic happens. There are funnels with flip-open covers that direct samples collected by the robotic arm to one of two instruments:

- The Chemistry and Mineralogy Experiment (CheMin), that uses X-rays to create diffusion patterns through soil and rock samples. The diffraction pattern identifies the minerals in the sample.
- The Sample Analysis at Mars instrument (SAM) houses a mass spectrometer, a gas chromatograph, and a tunable laser spectrometer. These instruments can spot organic compounds indicative of life or pre-biotic materials.

More small black-and-white cameras are mounted around the perimeter of the rover. These hazard-avoidance cameras, or Hazcams, help Curiosity to drive itself without hitting any obstacles. The rover also has a downward-facing camera designed to take continuous photographs of the descent, to help pinpoint the landing site and identify possible targets in nearby terrain of interest to the geologists.

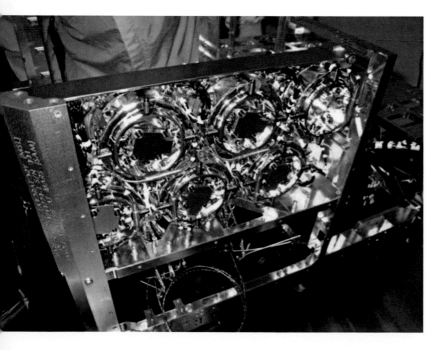

THE POWER OF CURIOSITY

Curiosity, as we shall call it from here on, had a long and painful gestation period. Everything seemed harder to do than foreseen, and adding international partners to the mix—while helping to share some costs—created delays and communication challenges. And there was one more major technical challenge that dragged down the schedule and chewed through the budget: landing.

When modern Mars landers of any kind reach the red planet, they do not go into orbit, looking for a nice landing spot, then descend when ready. For one thing, the fuel required to brake a spacecraft into orbit—especially something as heavy as Curiosity—is prohibitively heavy to launch. And we now have a sensational set of eyes orbiting the planet every day—MRO—and we know what is below already. So the preferred technique is simply to launch the spacecraft and aim for where Mars will be in six or seven months' time. You must then identify a precise spot right at the edge of the planet, skimming the thin layer of atmosphere, and head directly for the landing site. No wonder they called it Seven Minutes of Terror. The challenge with Curiosity was to figure out a way to rapidly decelerate all 2000lb (907kg) of Mars rover quickly and set it onto the surface at roughly walking pace. And while Mars' reduced gravity does help, at just 0.38 that of Earth, the atmosphere is not very helpful—it is just thick enough to cause frictional heating and blow the lander off-course, but not dense enough to slow the spacecraft as much as the engineers would have liked.

Given the successes of Pathfinder and the MERs, the natural impulse was to use the airbag approach and bounce to a landing. But the new rover was simultaneously both too heavy and too delicate for this method to succeed. So, what other landing mechanisms were in JPL's institutional memory? Curiosity was even heavier than the Viking landers at 2000lb (907kg) against about 1400lb (635kg)—but perhaps landing legs and rockets could work? This approach would require a landing stage, however, which would mean extra weight. It would also dictate that the rover would have to drive off the lander in order to gain access to the surface—and at this size and weight, that would be a complex task with many potential points-of-failure.

Lots of clever designs were suggested by the engineers... a crushable landing stage, that would collapse when it set down like a giant beercan, was proposed and studied, as were other designs. However, having a big, heavy rover atop a landing stage introduced another problem—it would be top-heavy. With most of the mass up high, it would be like balancing a bowling ball on a broomstick—unbalanced and prone to tipping, and that could spell disaster.

Then there were the requirements regarding the surface of Mars itself. The lander must be able to deal with coming down

OPPOSITE: The SAM instrument, or Surface Analysis at Mars, was a revolutionary device that crammed a room full of analytical instrumentation into a unit the size of a microwave oven.

RIGHT: Curiosity's parachutes were extensively tested at NASA's Ames Research Centerwind tunnel. These were the largest parachutes ever used at Mars, opening at supersonic speeds.

FAR RIGHT: Sky crane was the name given to the novel landing system created for Curiosity. The rover was too heavy for more traditional methods, so a unique combination of parachutes, rockets and winches were used to get the rover safely to the ground.

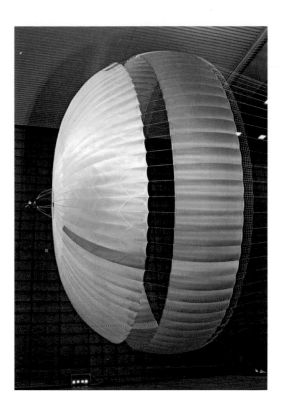

on top of rocks, as the Martian surface is covered with them, and there is no way to guarantee that the lander will not set down on a small boulder. You also do not want the rocket engines, which will be firing all the way down to the surface, to disturb the Martian soil any more than absolutely required. Finally, you want to make sure that you can drive right away, and will not waste valuable time trying to get clear of a giant landing stage and ramps.

In the end, the mission scientists started with a clean sheet of paper. The holdovers from past experience were a large heat shield and throttleable rocket engines, as used on the Viking landers, and a massive single parachute as used on all the landers. But nobody had ever built a heat shield this big, the know-how to build the Viking engines was long gone, and creating large parachutes that could support Curiosity's mass at supersonic speeds without tearing or snarling was a major challenge. In short, designing this mission could become a nightmare.

As they juggled all the various pieces, someone hit upon the notion of not using a landing stage at all—why not simply use the rover's wheels as the landing legs? They were strong, had the necessary clearance for rocks, and would already be dangling from the bottom of the rover anyway. And, since the rover is heavy, why not put the rockets up-top and let the rover hang from them—the complex mass-balancing problem would be solved. That left the issue of keeping the rockets from scorching and contaminating the ground; and that's when the sky crane concept was born.

The final design would work like this: the rover would enter the atmosphere encased in a protective aeroshell (as all the other landers). It would then glide, losing speed all the while, and at specific times would eject small but heavy "ballast weights" in order to shift its center of gravity and keep it in the proper orientation.

Then, the huge, supersonic parachute would deploy, slowing the spacecraft further. Once the 'chute had done its job, it would be cut loose, the heat shield would drop, and more rockets would fire to slow Curiosity to almost a hover. Then the rover would be lowered from the rocket stage via winches unspooling four nylon cords. When the rover signaled that it was on the ground, the cords would be cut and the still-firing rocket pack would fly away to crash in the distance. This would leave the rover on the ground, wheels extended, upright, and generally ready to go.

LIFT-OFF… AND LANDING?

As with the unconventional landing plan for Pathfinder, the sky crane concept was not an easy sell. But it was the least complicated way the engineers could figure out to get the heavy rover to the ground safely. They designed, built, and tested the rockets and the parachute, and started work on the massive 15ft/4.5m-wide heat shield. Each component had its share of problems, but one by one the problems were wrung-out and Curiosity was ready to go… almost.

Before anything can land on the surface of Mars, it must be sterilized to avoid contaminating the Martian surface or the sensitive

instrumentation within the rover with Earthly bacteria. A tenth of Viking's budget had gone toward its extensive decontamination process, a first in planetary exploration. While Curiosity was not held to as high a standard—it was not deemed necessary for this mission—this was still a complex process. The rover had a lot of sensitive electronics aboard, and the mission planners had to be careful not to overdo the cleaning process and damage anything.

After years of budget-busting tests, redesigns, and delays, Curiosity was finally ready to launch. On November 26, 2011, an Atlas V rocket, the largest available in the US inventory, thundered out of Cape Canaveral, headed for Mars.

Just over eight months later, on August 5, Curiosity encountered Mars. As always, the lander was on its own, due to the long radio delay. All landings on Mars are white-knucklers, but in this case, given the Curiosity's size and complexity, the bizarre landing system, and the costs—plus the fact that there was only one spacecraft this time—tensions were high. Enormous crowds gathered all over the US and in many other countries to watch the Seven Minutes of Terror on Jumbotron television screens. A few hundred members of the media were glued to the big screens at JPL's media center. And a choice few—those involved in the landing itself and some VIPs—were in JPL's mission control center.

The spacecraft encountered Mars' upper atmosphere at about 78 miles (125km) altitude, aimed for a landing zone only 12 miles (19km) long and in the belly of a crater, surrounded by peaks and valley walls (by comparison, the MER landing ellipse was 96 miles/154km long). As the heat shield absorbed the 3800°F (2093°C) heat, maneuvering jets kept it properly aligned for its long, sloping glide, as it slowed from 13,000 to 1000mph (20,921 to 1609km/h). All the while, the temperature inside the aeroshell stayed a comfortable 50°F (10°C).

Soon the tungsten ballast weights were ejected in a carefully timed sequence, tilting the spacecraft so that it would glide through the thin Martian air, shedding more speed. The parachute deployed as the heat shield dropped away, at an altitude of 7 miles (11km). The parachute gradually slowed Curiosity from 1000mph (1600km/h) to about 170mph (274km/h) before separating at just over 1 mile (1.6km) in altitude. The onboard radar kicked in to provide distance to ground measurements to the computer.

After a short free-fall, at just under 1 mile (1.6km) in altitude the rocket motors fired, steering it away from the still-descending parachute and aeroshell enclosure. At 800ft (244m) above Mars, still dropping at 30mph (48km/h), the wheel assembly was freed to drop and latch into place.

At 60ft (18m), the rockets slowed to almost a hover and Curiosity was winched down to the ground. The cables were cut and the descent stage flew off.

About fifteen minutes later, the radio signal was received at JPL. Al Chen, the communications officer for the landing, said in a slightly breathless voice, "Touchdown confirmed—we're safe on Mars!" The room erupted in cheers, and people all over the world clapped, laughed, shook their heads in disbelief, and cried. The greatest adventure of Mars exploration yet was underway.

ABOVE LEFT: After a white-knuckle Seven Minutes of Terror, flight controllers cheer the landing of Curiosity.

ABOVE RIGHT: The Mars Science Laboratory mission leaves Earth on an Atlas V rocket on November 26, 2011. It was already two years behind schedule.

OPPOSITE LEFT: After vetoing every other landing system previously used, engineers came up with the sky crane proposal. Here is one of the first sketches of the system eventually adopted, as drawn by engineer Tom Rivellini.

OPPOSITE RIGHT: The profile sketch of the sky crane shows how the rover would be suspended by nylon cords from the rocket pack, and lowered when the assembly was close enough to the ground. Many were surprised when it worked—perfectly.

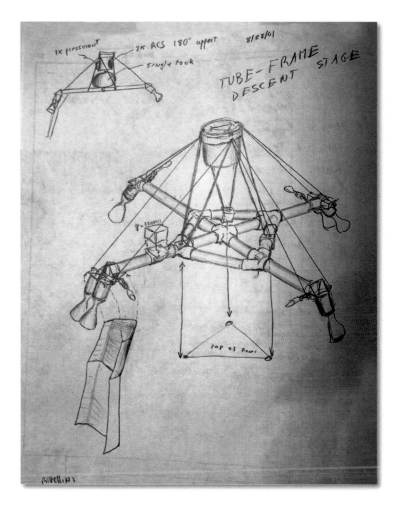

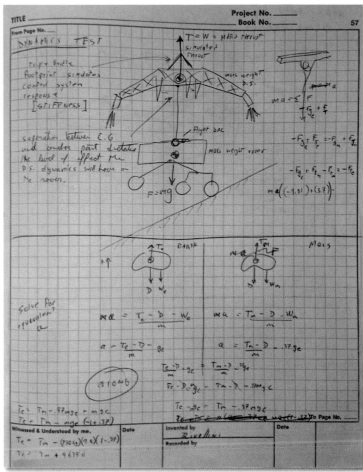

INSIDER'S VOICE

Tom Rivellini

Lead Mechanical Engineer for EDL Hardware

In the days leading to a final decision on how to design Curiosity's landing system, one last brainstorming session was held. Every idea was considered, no matter how outlandish.

"We spent three solid days evaluating every conceivable approach to landing a heavy rover on Mars. Miguell Sanmartin was eager to apply the lessons his team had learned on how to control the MER lander under its 20m [65ft] bridle. During the brainstorming sessions our thoughts kept revolving around simplifying egress. We found our way back into the insane notion of landing directly on the rover's wheels while suspended by Miguell's 'rocket on a rope.' I remember laughing at how nuts it was as we sketched it out on the white board. Once we got over the humor of its audacity we actually started liking the idea, a lot. It would answer nearly all of the problems we had been trying to solve…if it actually worked. The proof in the pudding came as we set out to try and 'break' the concept. We wanted to see if it worked past the whiteboard. Incredibly, each analysis showed that not only was this 'sky crane' not crazy, it was actually even better than we had thought it would be. We were hooked!"

WE FOUND WHAT WE CAME FOR

Curiosity arrived safely on Mars at 10.32pm Pacific Time on August 5, 2012. The first image returned was from a Hazcam, shot through the protective clear plastic cover. It was black-and-white and low resolution, but still a stunning look at the floor of Gale Crater.

GALE HAD BEEN SELECTED via an arduous process of scientists gathering, discussing their favorite candidate sites, and then arguing the pros and cons until they eliminated a few. They would then meet again a few months later to debate anew. In the end, there were a number of landing zones that met the criteria of interesting geology—enough different types of rock and terrain adjacent to each other, and within easy driving range, to qualify. Also of paramount importance was evidence of watery activity in the past. The geologists wanted an area that would net them lots of valuable samples to investigate, preferably rocks that would indicate past interaction with water. Something like an alluvial fan—an outlet of a water torrent where lots of different kinds of pebbles and soil would have been washed down and deposited during Mars' wet past—would be ideal.

However, the engineers had a say in this as well. The final selection would have to be sufficiently flat to give them a fighting chance for a safe landing, and allow the rover to drive relatively unimpeded across the surrounding landscape. Large crevasses, complicated and broken terrain, and sand dunes were off the table. The region also had to be sufficiently large to contain a reasonable landing ellipse that they could realistically target, and if it was at a lower altitude the denser atmosphere there would help them to land. These factors and others played into the decision making as well.

GALE CRATER

In the end, Gale Crater won the competition. The crater was almost 100 miles (160km) across, so the floor offered a large enough area to assure a safe landing. The images from Mars Odyssey and the Mars Reconnaissance Orbiter showed a generally flat crater floor that offered a large enough landing ellipse, and could be traversed with relative safety. There was plenty of complex geology at work there, and a vast alluvial fan was nearby, assuring a multitude of rock samples. Many of these would have come from the crater wall and possibly even the top of the rim. At the center of the crater was a large mountain, about 14,000ft (4267m) tall, that appeared to be the result of millennia of sedimentation. Examining the layers of the mountain would be like traveling through a time machine, offering a look at billions of years of geological history in one area. Finally, a number of differing types of geology converged not far from the likely touchdown spot, offering lots of opportunities for Curiosity to use a laser-powered spectrometer to analyze rocks, and its drill to deliver samples to its sophisticated onboard laboratory. So, Gale Crater was the choice.

Mount Sharp (also known as Aeolus Mons) was of particular interest. While it looks similar to other mountains in the center of craters—such as Tycho Crater on the moon—its formation could not have been more different. Tycho and many other craters in the solar system formed peaks in their center at the moment of impact, when a meteor or asteroid slammed into the moon or other body. The immense shock of such an impact will often form a "rebound" at the center, which can subsequently solidify into a sharp peak. This was not the case with Gale Crater.

Sometime in Mars' violent past, perhaps 3.8 billion years ago, Gale was formed by a large impact in an area known as Elysium Planum. The 96-mile (154-km) crater did not have a central peak when formed, or if it did, its remains are completely buried. Instead, over millions of years, the Gale was filled with wind and water-borne sediments—sand, dirt, and rock—from the nearby plains. Then, again over a very long span of time, those sediments were scoured out of the crater by incessant strong winds, leaving the central peak—Mount Sharp—behind. It was a heap of dust that solidified into a mountain—a haystack, as the geologists call it. The layering within the remaining mountain is a near-perfect record of billions of years of Martian history, with the oldest deposits at the bottom.

Curiosity set down between the base of Mount Sharp and the crater wall. The landing zone was named Bradbury Landing, after Ray Bradbury, the famed author of *The Martian Chronicles*, among other fictional works.

The rover sat for over a week while the engineers and controllers checked out its systems. Curiosity appeared to have survived its seven-minute landing well, though one weather sensor was damaged, apparently taken out of commission when it was struck by an errant pebble upon touchdown. However, there were two of these sensors, so losing one was a small price to pay for an otherwise perfect landing. It was time to go prospecting.

Instead of heading directly for the mountain, the scientists decided to drive in the opposite direction. There was an area nearby named Glenelg, which had three "geological units," or distinct rock types. It was felt that much could be learned by sampling material there. A note on names: previous lander and rover teams had invoked many whimsical monikers for rocks and regions near their machines, but this time, the geologists used names from famous (at least to geologists) regions of geological import on Earth. Glenelg was an area of interest to rock hounds in a region of Canada known as Yellowknife (a name which would also soon be affixed to a part of Gale Crater), and it seemed appropriate. Glenelg appeared to be layered bedrock, which would provide a good look at early Martian history, so they began their 0.25-mile (0.4km) drive to the region at a rate of about 20ft (6m) per Martian day.

Orbital images had shown that Glenelg featured "high thermal inertia," which meant that it retained heat well—better than sand or gravel would. This suggested that it might consist of cemented sediments, fine layering of deposits over time, and drilling there could provide a lot of valuable data. As the rover proceeded, the internal laboratories, SAM and CheMin, were tested while still empty. This provided baseline readings, and would identify any contaminants present that could later be subtracted from sample analysis, if necessary.

Along the way they came to an area of disrupted surface they named Hottah (again, after a similar feature on Earth). The view was electrifying. As John Grotzinger, the geologist in charge of the mission, put it, "[It] looked like someone jackhammered up a slab of city sidewalk…" Sticking out at an angle from the relatively flat valley floor, it did indeed look like a slab of concrete that had buckled under stress, and upon closer examination, all manner of pebbles and silt could be seen in layers. These pebbles were rounded and smooth, and cemented together with silt and sand—all clear indications of a long cycle of transport and deposition by running water, just like an ancient stream on Earth. Curiosity had, within weeks, encountered an ancient riverbed. This was the first clear indication of rapidly running water in Mars' past, and vast quantities of it. These rocks had been moved some distance, over a longer period of time, then had become cemented together after being deposited on the crater floor. Our visions of ancient Mars were getting wetter by the moment.

While thrilling, this was not, however, the best place to search for a habitable environment, nor to search for organic compounds, both primary goals of the mission. It was also not the best place to search for organic compounds. Over the next few weeks, the rover moved on to an area called Rocknest, where it was able to use the robotic arm to scoop up some samples, which it deposited into the CheMin instrument. The results were not Earth- (or Mars-) shaking—the analysis revealed that the dirt found in the sample was of basaltic, or volcanic, origins. It was not unlike what you might find in places on Earth such as the US Hawaiian islands, the result of ancient lava flows; it was also similar to the composition of Gusev Crater where Spirit had spent its time on Mars. Nonetheless, it was a good first run for the onboard laboratory. Combined with the observations at Hottah and its titled sedimentary strata, their

BELOW: About a month and a half after landing on Mars, Curiosity came upon this formation, which was named Hottah. It was an ancient riverbed, complete with rounded pebbles and layering.

NEXT PAGE: Two months into the mission, Curiosity reached Rocknest, seen here from the MastCam. The image is "white balanced," which makes the rocks appear the color they would appear on Earth.

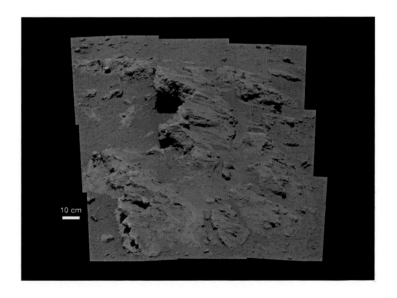

10 cm

hypotheses of the formation of Gale Crater, intuited from orbital observations, seemed solid. The idea that the bedrock had been formed from materials that spent a lot of time underwater, and the loose surface material had limited interaction with water, having blown down into the crater in later eras, appeared sound.

Curiosity made slow but steady progress as it headed into Yellowknife Bay, stopping to take more samples and do analytic runs with both the CheMin and SAM instruments. In late November of 2012, a story on NPR radio led to speculation that organic compounds, and possibly living organisms, had been found in the samples. This was rapidly picked up by secondary outlets and aroused much interest. In a major geological conference in December, speculations were laid to rest when the mission chief, Grotzinger, and the scientist in charge of the SAM instrument, Paul Mahaffy, gave detailed explanations of the investigative results. In brief, organics had been spotted in the samples, but whether they were of Martian origin or contaminants brought along from Earth was unclear. Grotzinger cautioned against drawing premature conclusions, counseling that "Patience is Curiosity's middle name." They were appreciative of the public's excitement, and would continue a thorough, careful search as the mission progressed.

HITTING PAYDIRT

In early 2013, a mission milestone was nigh: the first use of Curiosity's drill. This unit looks like a high-speed rotating screwdriver or chisel, and hammers as it spins to chip and grind rocks into powder for analysis. In mid-February, the geologists found a site they could agree on to take their first drill sample. It was called John Klein, and was in the Yellowknife Bay region, only about 200ft (61m) beyond Glenelg. Driving across Mars takes time.

Drilling was approached with great care—the first use of any instrument on a robotic explorer tens of millions of miles away can also be its last. After a full analysis of the drill site, the DRT wire brush tool scoured the rock, and ChemCam took laser-scorched shots to ascertain its basic make-up. Then the MAHLI camera took a close-up view. Everyone agreed that John Klein was sedimentary, riverbed-type rock, probably a mudstone, so it was time to drill.

There were concerns about the drill, however. It had exhibited intermittent short circuits in ground testing. The trouble-prone parts had been redesigned and replaced, but the violent nature of percussive drilling still had the engineers on edge. So care would be exercised in how often and how vigorously the drill was used. A couple of pilot holes were drilled, and then the final, sample-

ABOVE TOP: About six months into the mission, Curiosity captured this panoramic look at Yellowknife Bay over a series of days. The sun was just below the horizon in the lit-sky portion.

ABOVE: Curiosity takes a selfie near John Klein, its first drill site. This is accomplished by choreographing arm movements and snapping a series of images from the end of the arm, then assembling them to build an overall picture of the rover.

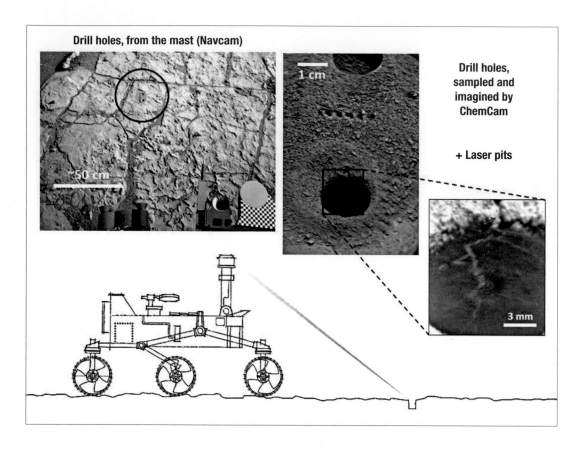

Drill holes, from the mast (Navcam)

~50 cm

1 cm

Drill holes, sampled and imagined by ChemCam

+ Laser pits

3 mm

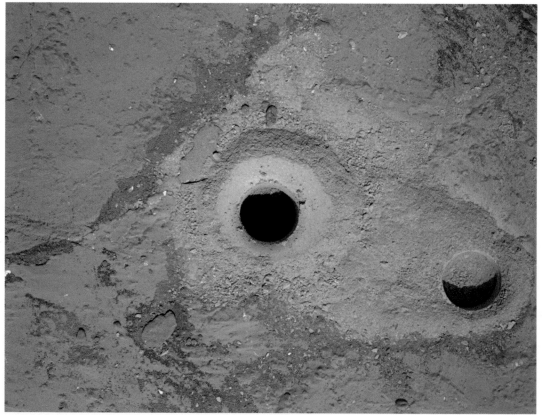

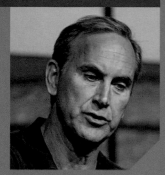

John Grotzinger

Mission Scientist, Mars
Science Laboratory

John Grotzinger was the mission scientist for MSL mission from its formative days through the end of the primary mission two years after landing. He remembers the August 5, 2012 landing well.

"It's this incredibly euphoric moment. But you're also sitting there and thinking, 'Man I sure hope this place delivers!' We spent a lot of time, six years, with the landing site selection."

With Curiosity's unique capabilities, making the proper choice for a landing site was critical. Dozens of candidate sites were painstakingly narrowed down to one: Gale Crater, with the majestic Mount Sharp rising from its center.

Gale seemed to have something for everyone... but other choices of landing sites had more to offer in specific areas. The problem with Gale Crater is that initially there wasn't this one thing that stood out and said 'pick me' other than Mount Sharp itself.

"Some people thought that Mount Sharp was just a pile of windblown dust. They thought we might go in there and not find anything. But the point is that this pile of windblown dust must have been altered with water, because we see clays, and we see the hydrated sulphates."

Barely six months after landing, when the first drill sample was analyzed, the wisdom of that selection was beyond reproach.

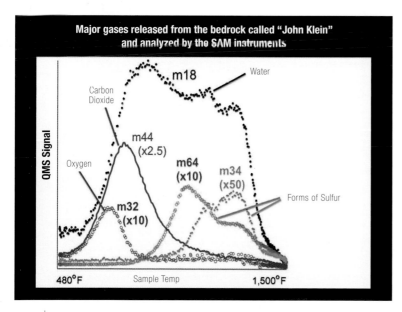

Major gases released from the bedrock called "John Klein" and analyzed by the SAM instruments

gathering hole was bored. The first surprise was immediate. While the face of the rock was red (as is most of the surface of Mars, due to the oxidized iron), the powder from the drilling was gray! This was of scientific import, showing that the oxidization on the surface did not go very deep—at least, not on this part of the planet.

Once the resulting gray powder was loaded into the receptacle on the robotic arm, a long process of manipulation delivered it to the funnel atop the rover for ingestion into the analytical instruments. As these analyses were being performed, the ChemCam laser zapped a series of spots inside the drill hole, which was just about the width of a dime, to provide a quick spectral analysis of the bore's walls.

After some delay due to balky computers, the results from SAM and CheMin came in. As the chief engineer, Rob Manning, said, "We found what we came for." Or as Grotzinger put it, "We hit paydirt." The analysis of this single, first drill sample by the onboard instruments represented mission success. The conclusion? Ancient Mars had been habitable. Microbes could have lived there. You could have drunk the standing water, which was in many places at least waist deep.

Curiosity had answered its primary query. The mission would continue, but everyone could breathe a bit easier: they had just "paid the bills," as Grotzinger put it, satisfying NASA's primary mission mandate in the first six months.

The rover quested onward, powered by its tireless nuclear power supply, with Mount Sharp looming in the distance.

OPPOSITE TOP: This graphic demonstrates how Curiosity uses the ChemCam laser to test the sites where it drills for samples. The circles in the top-left image show the test hole and sample hole. Top center are a series of laser "burns" between the holes. Top right are laser pits in the actual drill hole, which is about the diameter of a dime.

OPPOSITE BOTTOM: Drill holes at John Klein. To the right is the pilot, or test, hole, and in the center is the actual drill hole form which the sample was taken.

ABOVE: When the onboard instruments evaluated the tiny sample taken form the drill hole, the results were exciting: Mars had once been habitable, capable of supporting microbial life, if any were there.

ON TO MOUNT SHARP

Curiosity continued into 2013 with more drilling, soil analysis and ChemCam zapping that led to the stunning conclusion that not only had Mars been habitable at some point in the distant past, it may actually have been verdant. Moderate temperatures could have prevailed, there could have been mild, drinkable water in abundance, and the planet might have possessed "a cloudy atmosphere eloquent of fertility," as novelist H. G. Wells once wrote of Mars.

CONCURRENT WITH THESE SUCCESSES, mission controllers were experiencing recurring problems with the onboard computer and memory. Twin identical units (one backing up the other) ran Curiosity, using a microprocessor called a RAD 750. This chip was a military-grade version of the PowerPC 750, a chip introduced into the commercial market in 1997 by IBM and Motorola. By today's standards, it is an antique; it was already well into its second decade in 2012. However, militarized chips, which have been "hardened" against rough conditions and radiation, are always a few generations behind their civilian counterparts, and cost several hundred thousand dollars each. Nevertheless, even with this radiation proofing, the computers and associated flash memory were not immune to the degrading effects of incessant high-energy bombardment. Since the first few months of Curiosity's time on the surface of Mars, intermittent errors had occurred—likely due to stray cosmic ray particles striking the flash drives. The engineers had dealt with these in short order, but it was a continuing concern and each occurrence required shifting control of the rover from one onboard computer to the other while the affected system was diagnosed and reprogrammed.

For the first few months of 2013, the rover completed its survey of Yellowknife Bay. These investigations confirmed what it had learned at John Klein—billions of years ago, this part of Mars had been immersed in large amounts of standing and rushing water of benign chemistry that would have been a welcoming home for microbial life.

PRIMARY MISSION ACCOMPLISHED

By August, on its one-year anniversary, Curiosity had long-since passed its primary mission goals, driven 1 mile (1.6km) on the Martian surface, transmitted seventy thousand images back to Earth, and fired the ChemCam laser at two thousand targets. And the mission scientists were just getting started.

Subsequent drill samples continued to provide new evidence to refine their models for Gale Crater and, more generally, the entire planet. The crater floor was close to four billion years old, and for a long time much of it had been covered in water. Then the crater filled with dust and sedimentary deposits; still later that material was scoured away by wind over millions of years.

Organic materials were confirmed but their origins were uncertain. After exhaustive testing and retesting, the science team was convinced that these were not contaminants brought from Earth, which was a relief. Yet there were still questions—were the deposits of Martian origin, or did they come from a source outside, beyond Mars? Many meteors, of a type known as carbonaceous chondrites, include in their composition organic carbon. These were formed in the early solar system and continue to float around in abundance, and crash to the surface of moons and other worlds with regularity. Whether the organics found by Curiosity were

---

ABOVE: Curiosity's "brain" is a radiation-hardened (or "militarized") version of the same PowerPC chip that powered a previous generation of Macintosh computers. Two identical units are used to assure redundancy.

OPPOSITE: This graph shows radiation measurements taken for the first 300 days on Mars. The readings indicate that properly shielded humans could survive on the planet; the only major spike is from a solar event to the right about day 245.

Ashwin Vasavada

Project Scientist, Mars Science Laboratory

"After nearly four years on Mars, Curiosity is doing well, but showing a few signs of aging. In all likelihood we're halfway through the mission, giving me a great sense of pride in all our team and rover have accomplished, but also a sense of urgency to do all we can in the future.

"We'd really like to get high enough on the mountain to reach a layer of rock that appears in orbital data to be enriched in sulfate salts, perhaps because it formed in a more arid environment. The transition from freshwater lakes to salty, evaporating pools could reveal the story of how Mars changed as a planet. About 3.6 billion years ago when Gale crater formed, Mars was quite different than the planet we are exploring today with Curiosity. The atmosphere was thicker then, warming the climate and providing stability for liquid water.

"Over the next few years, Curiosity will ascend through the layered rock of Mount Sharp, answering questions we have about the crater and the long-lasting lakes that were an integral part of Mount Sharp's formation: just how long did these lakes last? Was the climatic shift to today's cold and dry conditions gradual, or abrupt? Most importantly, how long did Mars have an environment that offered a potential habitat for life?"

from one of these meteorites, or a source indigenous to Mars, was uncertain. And since Curiosity is not a life science mission per se, there was no direct way to test whether or not the source was biological. But in any case, it was a promising development, and added to the already long list of accomplishments for the mission.

Any analysis of Martian surface samples is complicated by the presence of perchlorate—a nasty chemical that permeates Martian soil—as well as the fact that any exposed surface on Mars has been irradiated by copious amounts of ultraviolet radiation

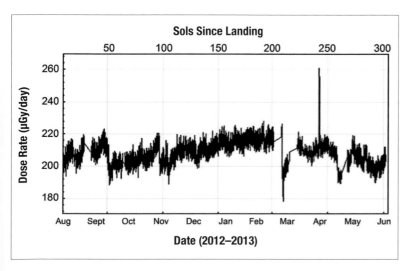

from the sun for millions of years. Drilling helped to get past the worst of it, but any future attempts would be well served by the ability to drill far deeper—at least 3ft (1m)—and take core samples for analysis. Anything living closer to the surface is likely to have been killed-off by massive doses of both solar radiation and cosmic rays from deeper in space. Exceptions might be areas protected by overhanging cliffs or soil underneath rocks, and these are both options for future studies, possibly with the 2020 rover.

Reaching the one-year mark for Curiosity also provided a convenient baseline measurement from the RAD radiation sensor, the instrument measuring the radiation exposure that human explorers would suffer during a stay of over twelve months on Mars. When the numbers were scrutinized, it appeared that a human who spent 2.4 years on a Mars mission—180 days traveling to Mars, 500 on the surface and 180 days back—would be dosed with about one sievert of radiation (this assumes minimal shielding). This would represent about a five percent increase in the risk of developing cancer over the astronaut's subsequent lifetime. Adequate shielding from radiation during the transit to Mars, and while on its surface, will have to be a priority for any future explorers.

That the ancient environment had once been habitable was now beyond doubt, at least in the Gale Crater neighborhood. The water present in the crater some 3.5–4 billion years before had been low in salinity ("You could drink it," said the mission scientists). It also

had a neutral pH and contained chemicals—iron and sulfur—that microbes could use for nutrition.

It was time to head for higher ground. After a series of other stops and investigations, Curiosity turned finally toward Mount Sharp. The rover would have to cross some difficult territory along the way—sandy dunes and rough, rocky terrain—but once it got started, progress was steady, if slow. The frightening moments from the MER Spirit's mission (see pages 94–101), in which it repeatedly (and finally, fatally) got stuck in sand and drift, were not an issue for Curiosity, due in part to its larger size and bigger wheels. But another issue related to its transport across the Martian terrain began to dog its progress: damage to those large wheels.

Curiosity's wheels had been designed to be the perfect compromise between traction, strength, and lightness. Machined from a single block of aluminum, the wheels are roughly the size and shape of a small beer keg. The surface metal is *extremely* thin, with much stronger traction cleats (called "grousers") criss-crossing the structure. Each end of the wheel is wrapped inward to add strength, and in the centerline of the interior is a thick radial rim.

Spokes attach this rim to the hub. They look tough, but are actually, as it turns out, rather delicate.

This became clear when the punctures and tears began to show. First a few small holes, then larger tears, some the length and width of a thumb, showed up all over the wheels. These appeared only in the thin foil between the grousers (cleats), and did not affect the strength or shape of the wheel. Nonetheless, it caught the engineers' attention. There are holes machined in the wheels to allow for the numbers of rotations to be tracked in the dirt, but these are small and in single rows. What was showing up now were longer, random and looked like they might imperil the wheels and, thus, the mission.

Tests at JPL demonstrated that in certain circumstances it was quite easy to puncture the wheel face. While the strength of the wheel overall was not much diminished by a few rips and punctures, it was still something the mission scientists wanted to avoid. And, as luck would have it, Curiosity had just entered an area that was covered with the worst kind of rocks they could imagine—they looked just like dragon's teeth. An otherwise flat plain was littered

ABOVE TOP: Dingo Gap contained a sand dune that Curiosity had to cross to move from Yellowknife Bay on its trek to Mount Sharp. The rover carefully tested traction before initiating the slow and careful crossing.

ABOVE: These wheels represent three generations of Mars rovers. The smallest, center, is from Pathfinder's Sojourner rover; left is the MER rover's wheel, and right is the massive wheel of Curiosity.

RIGHT: This is one example of the many tears and punctures suffered by Curiosity's wheels on its travels across the floor of Gale Crater. While the metal on the wheel surface is thin and easily damaged, the cleats, known as grousers, are much stronger, and reinforced by metal rims inside the wheel.

ON TO MOUNT SHARP

ABOVE: In May 2015, Curiosity reached Maria's Pass where an older area of mudstone—the pale band across the center—is overlain by a more recent sandstone formation.

RIGHT: Curiosity, coated with red dust, takes a selfie at Windjana. A drill sample of sandstone rock was taken near here.

with sharp, jagged spikes left behind by bizarre erosional patterns, in which the softer soil had blown away over the millennia, leaving behind these "teeth." Curiosity had to cross them regardless of risk—it was stuck with no way out save forward.

The scientists picked the safest course they could find and even spent time driving backward, which, due to the design of the suspension, put less pressure on the leading wheels. Some rocks that would puncture a wheel when driving forward merely dented them in reverse. Once clear of the dragon's teeth, the rover drivers were much more careful in route selection, at one point driving up the sandy center of an old riverbed for hundreds of yards. It was the smoothest, safest surface they could find, and the tactic worked.

A MARTIAN LANDMARK

In late June 2014, the MSL team celebrated its first Martian year, 687 Earth days—the duration of its primary mission. Funding was ultimately extended for another two years, and Mount Sharp was in the crosshairs. More drilling, sampling, and analysis was conducted as they approached the slopes of the mountain… then, in December, NASA announced that they had detected a surprising rise in methane in the vicinity of the rover. It was a brief event, unprecedented in the mission and puzzling to all concerned. The story of methane and Mars is, as you know, both compelling and confusing. There was no way to tell where this methane plume had come from—only that it was localized around the rover for a brief time. The source could have been geological or biological. The observation has not reoccurred, but everyone involved has their eyes peeled for another methane boom.

New drilling indicated organics again, and other data from the samples helped to pin down the timing and rate of water loss at Gale Crater. In early 2015, nitrogen was detected by the SAM instrument, and was a further indicator of the once habitable nature of the planet.

Now, items on the menu for any hungry bacteria in the distant past included: nitrogen, sulfur, hydrogen, oxygen, phosphorus, and carbon. And of course sunlight, though any Martian microbes could have been chemolithotrophs—microbes that live in, and eat, rock.

As 2015 progressed, the rover began to head into the foothills of the mountain. There is a very gentle, sloping approach to the valleys that cut through the base of Mount Sharp that Curiosity is using to slowly gain altitude. The data coming back from the rover indicates marked changes in the strata as it drives upslope, with much more silica present than is contained in the basalts nearer the crater floor—in some areas, the rocks consist of almost ninety percent silica. This is indicative of interaction with water, but what it means in a broader context, and why the concentrations are so high in these particular layers of rock, are still open questions.

As 2015 drew to a close, key members of the Mars Science Laboratory team released a major scientific paper—it was just one of dozens that had been written about the mission, but was unique in that it summarized the sweep of operations since Curiosity's arrival at Mars.

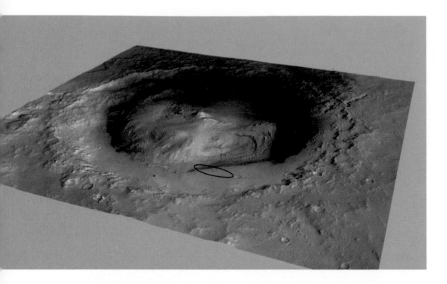

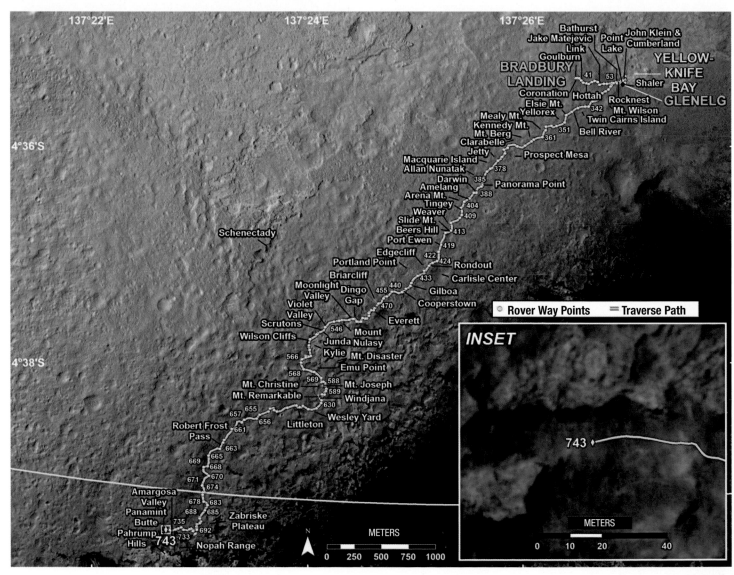

137°22'E   137°24'E   137°26'E

Bathurst
Jake Matejevic   Point   John Klein &
Link   Lake   Cumberland
Goulburn   YELLOW-
BRADBURY   41   53   KNIFE
LANDING   Shaler   BAY
Coronation   Hottah   GLENELG
Elsie Mt.   342   Mt. Wilson
Yellorex   Rocknest
Mealy Mt.   Twin Cairns Island
Kennedy Mt.   351
Mt. Berg   361   Bell River
Clarabelle
Jetty   Prospect Mesa
Macquarie Island   378
Allan Nunatak
Darwin   385
Amelang   Panorama Point
Arena Mt.   388
Tingey   404
Weaver   409
Slide Mt.
Beers Hill   413
Port Ewen
     419
Edgecliff   422
Portland Point   424   Rondout
Briarcliff   433   Carlisle Center
Moonlight   Dingo   455   440   Gilboa
Valley   Gap   470   Cooperstown
Violet
Valley   Everett
Scrutons
Wilson Cliffs   546   Mount
    Junda   Nulasy
   566   Kylie   Mt. Disaster
   568   Emu Point
    569   588   Mt. Joseph
Mt. Christine   589   Windjana
Mt. Remarkable   630
   657   655   Wesley Yard
Robert Frost   661   656   Littleton
Pass   663
   669   665
   668
   671   670
    674
Amargosa   678   683
Valley   688   685
Panamint   735   Zabriske
Butte   692   Plateau
Pahrump   733
Hills   **743**   Nopah Range

Schenectady

○ Rover Way Points   = Traverse Path

4°36'S

4°38'S

N

METERS
0   250   500   750   1000

INSET

743

METERS
0   10   20   40

The authors confirmed that Gale Crater had been filled by sediments, that long-lived streams and lakes were active inside between 3.3 to 3.8 billion years ago, and that ancient Mars more closely resembled Earth than the Mars of today. The lower portions of Gale filled in during a span of 400–500 million years, much faster than previous estimates. This sedimentation is observed to be as high as 600–700ft (183–213m), based on orbiter images. In all, the sedimentation in Gale Crater could extend as high as 0.5 mile (0.8km), with the layers above 700ft (213m) likely deposited by wind rather than water. It is a remarkable height for just 500 million years of activity.

The paper was published in the same month that NASA announced the discovery of small, seasonal flows of extremely salty water that are still active on Mars. These appear in warmer weather and vanish quickly, vaporizing into the thin air. However, in the case of the ancient streams, lakes, and rivers found by Curiosity, it is a mystery how these large bodies of water managed to exist. The geological evidence points to a denser, wetter, warmer atmosphere—but the climate models for ancient times do not add up to what the geologists are seeing. It is back to the drawing board for the ancient Mars weather theorists.

Perhaps John Grotzinger best summed up the mission when discussing the most recent discoveries: "We have tended to think of Mars as being simple," he said. "We once thought of the Earth as being simple, too. But the more you look into it, questions come up because you're beginning to fathom the real complexity of what we see on Mars. This is a good time to go back to re-evaluate all our assumptions. Something is missing somewhere."

And that is the perfect reason for the ongoing exploration of Mars. The ExoMars orbiter/lander will arrive in 2016, the ExoMars rover and NASA's InSight orbiter in 2018. The Mars 2020 rover should launch in 2022. Between them, these four advanced spacecraft may just find what is missing.

OPPOSITE TOP LEFT: Computer-generated image of Gale Crater as it looks today, with MSL's landing ellipse superimposed on the crater floor.

OPPOSITE TOP RIGHT: This image, created using topographic data of Gale Crater, shows how the region might have looked 3.5 billion years ago when filled with liquid water.

OPPOSITE BOTTOM: A traverse map of Curiosity's journey from landing until mid-2015. The final point of the traverse indicated here is called Pahrump Hills (seen in inset), which contained some of the oldest layers of rock in Gale Crater.

ABOVE: This image from September 2015, three years into Curiosity's mission, looks towards the high elevations of Mount Sharp. The mountain in the foreground is about 2 miles (3 km) away, and contains large amounts of hematite, a mineral formed in the presence of water.

BELOW: The beginning of the climb: the foothills of Mount Sharp. This area is where Curiosity began its long climb to higher elevations of the central peak in Gale Crater, and where it will find increasingly younger layers of rock. Many questions about the formation of Gale, and the role of water in Mars' past, will be answered here.

~2 KM

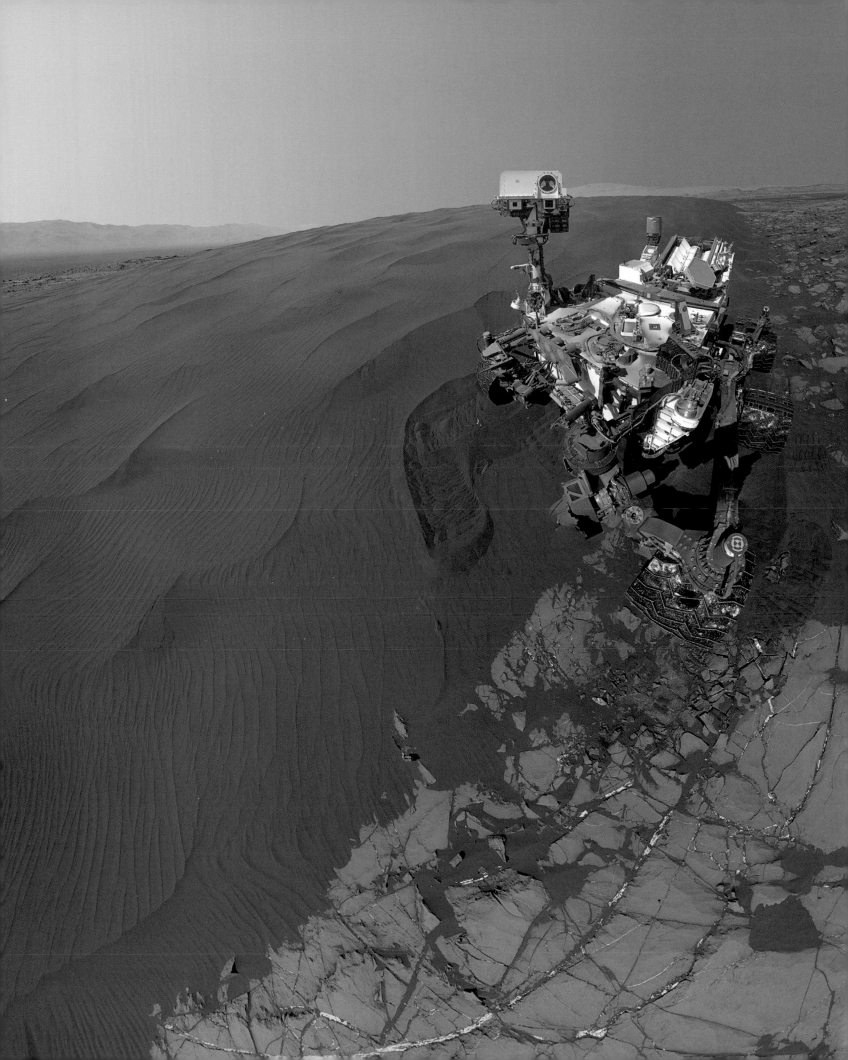

OPPOSITE: In early 2016, Curiosity reached Namib Dune, where the rover tested the sand by scuffing it with a wheel and ingesting a sample for onboard analysis.

ABOVE: In March 2016, Curiosity reached this area of old, weathered rock called Sesrium Canyon. The drive there was interrupted by a recurring short-circuit in its nuclear power supply that slowed but has not stopped the rover.

YIN AND YANG: MAVEN AND MANGALYAAN

For decades, the exploration of Mars was strictly a two-nation undertaking, with only the US having enjoyed success. While the attrition rate of Mars-bound spacecraft had been nearly half, the vast majority of failures involved Soviet missions. Japan attempted their own small orbiter in 1998, but their mission failed as well. Mars was, in effect, US territory.

IN 2003, EUROPE'S MARS EXPRESS successfully orbited Mars (see page 84), making Europe only the second power to achieve this feat, and on a maiden flight. Russia's most recent attempt, launched in 2011, was called Fobos-Grunt, and carried a Chinese orbiter along with their own, upping the list of nations that have attempted Mars missions to five. However, the spacecraft failed to leave Earth orbit, burning up in an uncontrolled re-entry in 2012. Meanwhile, NASA's missions continued to head off to Mars every few years.

But the tables tilted in 2013. Forty-eight years after the first successful Mars flyby by Mariner 4 (see page 23), there was a new player in the game. India launched their Mangalyaan spacecraft, also known as the Mars Orbiter Mission, in November 2013 and successfully orbited the planet on September 24, 2014.

INDIAN EXPLOITS

This was no small, experimental orbiter launched on a borrowed rocket, either. Mangalyaan tipped the scales at almost 3000lb (1360kg) and was launched on India's own proven PSLV rocket. As the fourth country to reach the red planet, and the first Asian nation to do so, the success was a moment of intense and well-deserved pride. NASA backstopped the mission with advisory and tracking and navigation assistance, and continues to lend its powerful Deep Space Network to the effort, but the design, assembly, and launch of the mission were all out of India. The country has also

built its own deep space network for communicating with the spacecraft. The sum of this mission design was a stunningly low overall cost, estimated at just over $70 million—a true bargain in planetary exploration.

The spacecraft design was based on that of an earlier Indian mission, the Chandrayaan lunar orbiter, which carried out successful investigations of the moon in 2008. In that same year, initial announcements were made indicating the country's intentions to reach next for Mars, and the mission was approved in 2012 for a launch date in 2013—which was not a lot of time to build a first Mars orbiter. This was also a new record for fast-tracking an interplanetary spacecraft.

While in many ways this was a technology demonstration mission, the Indians' scientific objectives were reasonably ambitious. Along with the usual investigation of the surface of the planet, Mangalyaan would also search for methane in the Martian atmosphere, as well as exploring the upper atmospheric dynamics and mechanisms behind the continuing loss of Martian atmosphere into space. Instrumentation included: a Lyman-Alpha Photometer (LAP) to investigate water loss into space; a Methane Sensor for Mars (MSM); the Mars Exospheric Neutral Composition Analyzer (MENCA); a mass spectrometer which would also seek methane among other objectives; a Thermal Infrared Imaging Spectrometer (TIS) to measure surface temperatures, critical for researching

OPPOSITE TOP: Controllers monitor Mangalyaan, the Mars Orbiter Mission, from the India Space Research Organization's control center.

OPPOSITE BOTTOM: Electromagnetic interference tests are conducted in an anechoic chamber, which absorbs all abound and electromagnetic waves.

MANGALYAAN

MISSION TYPE: Orbiter
LAUNCH DATE: November 5, 2013
LAUNCH VEHICLE: Polar Satellite
Launch Vehicle
ARRIVAL DATE: September 24, 2014
MISSION TERMINATION:
Mission ongoing
MISSION DURATION: ~1.5 years, six
months + multiple extensions
SPACECRAFT MASS: 2948lb
(1337kg)

mineral distribution; and the Mars Color Camera (MCC), for imaging.

After completing its initial six-month mission, Mangalyaan was approved to continue into the indefinite future. Plenty of maneuvering fuel remains in the spacecraft for extended operations. The spacecraft has been successful in tracking concentrations of dust in the atmosphere, as well as atmospheric compositions at various altitudes. The Martian moon Deimos was also studied, with rare backside views sent home by the onboard camera. Since Deimos is tidally locked to Mars, meaning that the same side always faces the planet, the fact that the far side was much smoother and of a different color tone than the one facing the planet was interesting, prompting suggestions for future investigation.

MAVEN'S WAY

Just weeks after India's launch, NASA sent its long-anticipated MAVEN spacecraft to Mars. MAVEN is an acronym for Mars Atmosphere and Evolved Evolution, and was designed to study how the atmosphere of the planet has been largely lost to space. Maven also arrived in September of 2014, two days ahead of Mangalyaan, on the 22nd. This was one of the last Scout-class missions, joining the Phoenix lander in that distinction. MAVEN's arrival at Mars at the same time as that of Mangalyaan was a bit of yin to the Indian spacecraft's yang: a highly specialized Mars probe from a fifty-year veteran, NASA, meets the newest space power's generalist, the technology-proving maiden flight of the Indian planetary exploration program.

The evolution of Mars' atmosphere is a puzzle. Surface rovers indicate a once verdant planet with large bodies of standing water in the distant past; recent estimates suggest about three and a half to four billion years ago. That determination came from the geologists on the MER and Curiosity missions, backing up decades of speculation driven by the water-caused erosion patterns that are visible in orbital images from Mariner 9 onward. However, other scientists, the ones who work on ancient Martian weather and climate (they actually have a name: paleoclimatologists), do not see indications of the dense and warm ancient atmosphere that would be required to keep that water from evaporating rapidly, or freezing instantly, as it would today. So where did it go? The causes and mechanisms of this extreme shift—it is estimated that Mars has lost up to ninety-nine percent of its ancient atmosphere—are MAVEN's primary mission.

MAVEN's design is derived from those of the Mars Odyssey and Mars Reconnaissance orbiters, and the spacecraft was built by the same US company, Lockheed Martin. The science package is a bit different from previous orbiters, as MAVEN's mission is very specific. Everything onboard is designed for investigation of the Martian atmosphere, and the

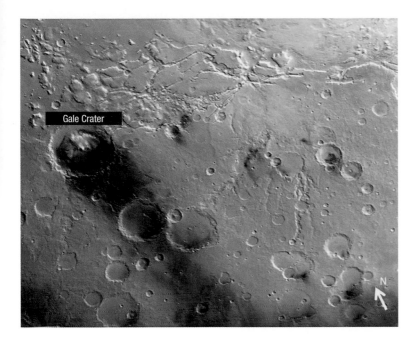

Gale Crater

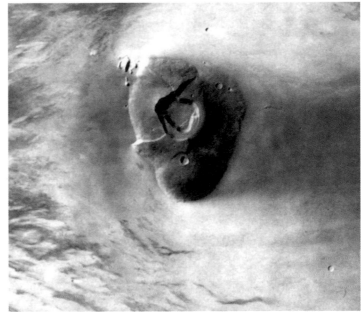

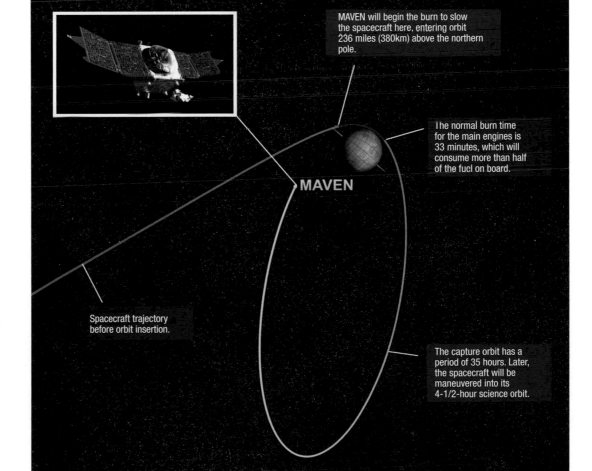

MAVEN will begin the burn to slow the spacecraft here, entering orbit 236 miles (380km) above the northern pole.

The normal burn time for the main engines is 33 minutes, which will consume more than half of the fuel on board.

MAVEN

Spacecraft trajectory before orbit insertion.

The capture orbit has a period of 35 hours. Later, the spacecraft will be maneuvered into its 4-1/2-hour science orbit.

OPPOSITE: MAVEN mission logo, 2013.

ABOVE LEFT: Gale Crater, Curiosity's home base, as imaged by Mangalyaan's Mars Color Camera.

ABOVE RIGHT: The massive Martian volcano Tharsis Tholus, imaged by Mangalyaan. The volcano measures 96 miles (155km) at its widest, and was first discovered by Mariner 9.

RIGHT: A trajectory plot showing how NASA's MAVEN orbiter reached Mars. As it passed the planet, it burned half its fuel load to brake itself into the long, looping orbit seen here. Subsequent rocket firings put MAVEN into its final orbit. While the spacecraft did not use aerobraking as previous orbiters had, it has dipped low into the atmosphere to conduct its science mission.

spacecraft does not carry a camera. MAVEN has: instruments for analyzing the solar wind and its interaction with the Martian atmosphere; a magnetometer for measuring Mars' weak magnetic field (thought to be a primary culprit in atmospheric loss over time); an Imaging Ultraviolet Spectrometer (IUVS) for measuring the upper layers of the atmosphere; and a Neutral Gas and Ion Mass Spectrometer to examine atmospheric gas make-up. This final instrument is similar to Mangalyaan's MENCA device.

MAVEN may be a budget-constrained mission, but it is no lightweight. The main body is a cube about 8ft (2.4m) to a side, and when the solar panels are extended it is as much as 38ft (11.5m) wide. The spacecraft weighs in at a hefty 5400lb (2449kg).

After MAVEN's first year of observations, researchers concluded that the solar wind, the continual bombardment of the inner planets due to energetic particles emitted from the sun, had in fact stripped away most of the Martian atmosphere due to the planet's lack of a magnetic field. This was as expected. They also discovered that the rate of atmospheric depletion increases during solar storms, when energy levels coming from the sun are higher. The rate of loss was charted at about a 0.25lb (113g) per second, which does not sound like a lot for an entire planet, but over billions of years, it adds up. And since atmospheric loss accelerates when the sun is more active, which is exactly how the sun behaved earlier in its history, Mars has taken quite a battering over the eons. The atmosphere has suffered dramatically, with only a scant one percent of the original dense envelope remaining.

When the high-energy particles streaming from the sun, which are composed primarily of protons and electrons, slam into Mars at speeds of up to 1 million mph (1.6 million km/h), it causes an

OPPOSITE: MAVEN launched on November 18, 2013, on an Atlas V rocket from Cape Canaveral, Florida.

ABOVE: MAVEN's size is apparent when compared to the technicians preparing it for flight.

BELOW: This graphic demonstrates how Mars lost its atmosphere, one of MAVEN's primary research goals. Each image shows the escape of different atmospheric elements over time due to solar energy hitting Mars. Unlike Earth, Mars is unprotected by a magnetic field.

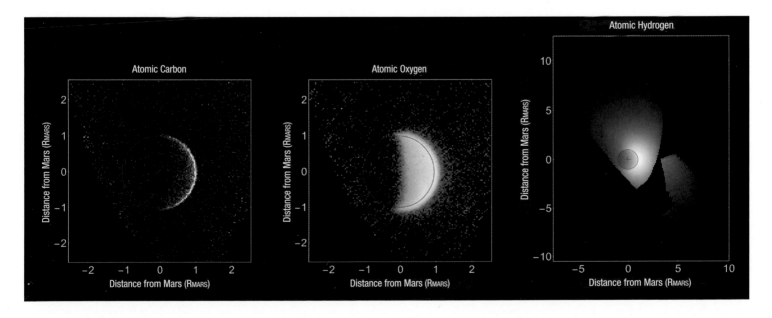

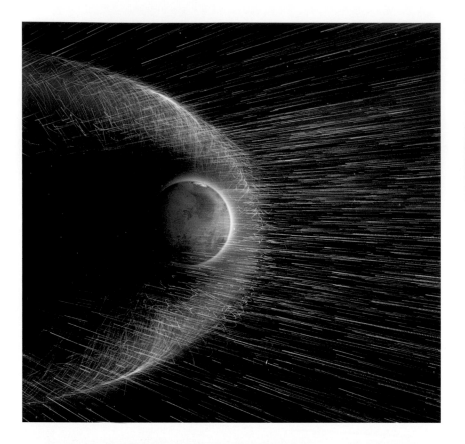

ABOVE LEFT: Artistic depiction of how the solar wind, massive amounts of energetic particles from the sun, sweeps past Mars, scouring much of its atmosphere. Mars, in contrast to earth, has a very weak magnetic field.

LEFT: In the case of Earth, the solar wind is channeled and redirected by our planet's much more active magnetosphere.

ABOVE RIGHT: Artist's conception of a solar storm, a time of greatly increased energy, and how it scours ions away from the Martian atmosphere.

MAVEN

MISSION TYPE: Mars orbiter
LAUNCH DATE: November 18, 2013
LAUNCH VEHICLE: Atlas V
ARRIVAL DATE: September 22, 2014
MISSION TERMINATION: Mission ongoing
MISSION DURATION: 2+ years
SPACECRAFT MASS: 1784lb (809kg)

electrical field to form, much like an electrical generator. This in turn accelerates ions that are present in Mars' atmosphere, causing them to fly away from the planet and into space. This has been going on for a very long time—billions of years—and the lengthy process has resulted in a cold, dry planet.

MAVEN continues its investigations into more detailed questions about atmospheric loss, but the big questions about the thinning of Mars' atmosphere—from when it was wet and potentially habitable through today's wispy remnant—seem to have been answered. MAVEN's mission extensions will continue to fill in the remaining blanks, as will India's next venture, tentatively scheduled to land on Mars in 2018.

NEXT STEPS: INSIGHT AND THE MARS 2020 ROVER

With prime launch opportunities occurring every two years, the pressure is on to keep the spacecraft flying to Mars. However, with budgets constrained and mission timetable slips, these windows of opportunity are not always easy to accommodate. With MAVEN and Mangalyaan both reaching Mars in 2014, NASA hoped to launch a new lander—named InSight—for arrival in 2016. This was a Discovery-class mission with a budget cap of $425 million. Consequently, to save time and money, the general design from the Mars Phoenix lander (*see* page 116) was adapted to fit the mission—to probe deep within the planet and find out what makes Mars tick.

A NEW INSIGHT INTO MARS...

InSight, like MAVEN, was a specialty spacecraft created to follow a specific course of investigation: to gather data about the overall structure of Mars by examining its geological interior. Mars is one of the terrestrial planets of our solar system, rocky worlds that also include Mercury, Venus, and Earth. The rest of the planets are gaseous, and followed very different evolutionary paths. Much has already been learned—not just about Mars, but about rocky planets in general—by studying the surface of Mars, which has generally remained unchanged since ancient times. It has been sculpted by wind and sand, and in much earlier times water, but compared to the Earth—which has endured billions of years of severe weathering and violent structural change—Mars is like a geological museum.

The interior structure and dynamics of Mars is one of the last major parts of the planet that remain completely unexplored. By studying the hot core of the planet (which is much smaller than that of Earth), its mantle, and crust, and comparing these results to what we know of the Earth, much should be gleaned about the early formation of all the rocky worlds.

The 800-lb (363-kg) InSight lander carries a very specialized scientific package to accomplish its goals, with two investigative instruments topping the list. One is the Seismic Experiment for Interior Structure (SEIS), a seismometer capable of measuring Marsquakes as well as reading the interior structure of the planet. It is so sensitive that it can even detect remote meteor impacts. The Viking landers carried basic seismometers (only one of which worked), but the SEIS unit is a much more sensitive device. The second primary instrument is the Heat flow and Physical Properties Package (HP3), which will measure heat coming to the surface from the core of Mars.

Other included experiments are the Rotation and Interior Structure Experiment, which uses the spacecraft's radio signal to more exactly measure the planet's rotation. This will help pin down internal structure and mass. Two small cameras—one on the lander's "deck" and another on the arm—will aid in positioning and observing the placement of the SEIS seismic probe. A magnetometer and weather station round out the lander's instrumentation.

The HP3 heat flow experiment will be a first on Mars, and is a long sought after experiment for the red planet. The Apollo missions deployed similar instruments on the moon in the 1970s, but had astronauts to set up the device by drilling sensor-tipped stakes deep into the lunar surface. This will be the first time a similar experiment had been attempted robotically—the 18-in (45-cm) spike will burrow up to 16ft (5m) into the Martian soil with an internal hammering mechanism. A cable trailing behind it contains heat sensors positioned at regular intervals. By measuring the rate at which heat radiating from deep within the planet reaches the staggered temperature receptors, scientists can extrapolate

OPPOSITE: The InSIGHT lander, now set to launch in 2018. The important components are the WT/SEIS, the seismometer sensor, and the HP3, the burrowing heat flow probe. These two devices are at the heart of InSIGHT's mission: to explore the core of Mars.

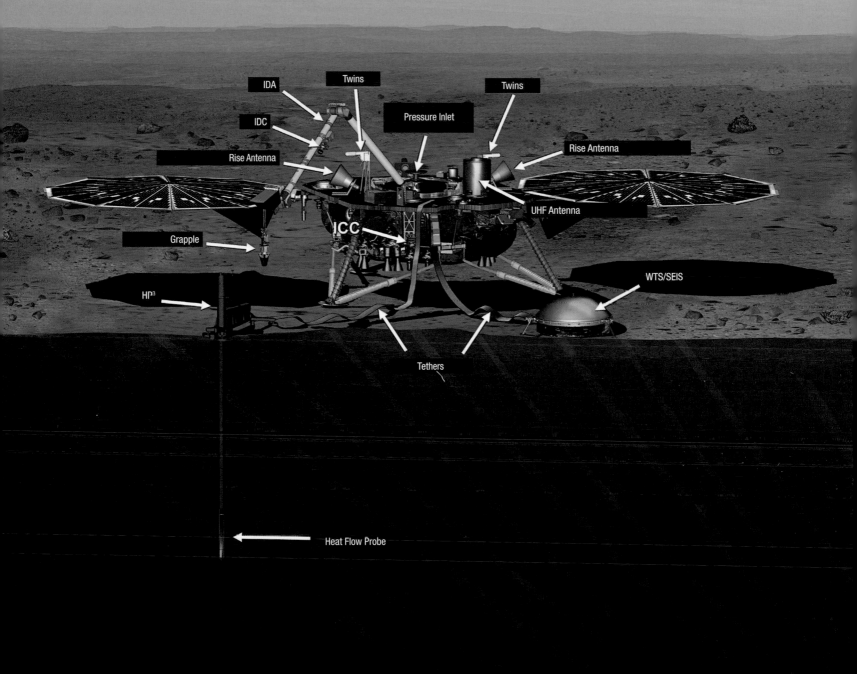

much about the structure of the planet's core, as well as the local geological make-up of the region.

After landing, the SEIS's seismic sensor must be grappled by InSight's robotic arm, offloaded, and set onto the surface—another first for a Mars lander. While fully tested and seemingly simple, every "first" in a planetary robotic mission must be simulated and studied on Earth until all the possible failure scenarios are understood and worked out. The heat probe and seismic sensor teams at JPL and elsewhere will be sweating until their instruments are successfully positioned and safely sending home data.

A new communication technology for Mars spacecraft will be tested during the InSight mission: two tiny cubesats—small, square Martian satellites—will be deployed in Martian orbit before InSight lands. These mini-orbiters will serve as dedicated data-relay units, and their primary purpose is to send back the radio signals from the InSight lander as it descends to the Martian surface. If they are effective, more cubesats will likely be sent into Martian orbit, providing landing support and supplementing the larger scientific spacecraft such as Mars Odyssey and the Mars Reconnaissance Orbiter in their secondary roles as radio relay stations for the landers and rovers of the future.

The InSight spacecraft was built, tested, and prepared for a 2016 launch, but problems with the seismometer caused a delay. The unit is vacuum-sealed to allow the extremely sensitive detection devices inside to function in an almost frictionless environment. In pre-flight testing, the vacuum seal was found to be defective, so JPL will attempt to fix or replace the unit. For a time the entire mission appeared to be in jeopardy, but a new launch date in 2018 was recently announced.

THE MARS 2020 ROVER

The Mars 2020 rover is the next flagship mission to the red planet, rivaling Curiosity (see page 124) in complexity, ambition, and budget. The machine has been designed on the same chassis as the Curiosity rover in order to save money, and the mission should come in at far less cost—a total of roughly $1.5 billion, as opposed to the $2.5 billion spent on Curiosity. Mars 2020 will carry a different and even more complex set of instruments within, and is set to launch in its namesake year, 2020.

The objectives for the 2020 rover are impressive. Its primary goals will be to continue the search for the past and present habitability of Mars, continuing Curiosity's mission with greater

OPPOSITE: InSIGHT lands on Mars, artist's conception. InSIGHT is based on the design of the successful Mars Phoenix lander.

BELOW LEFT: Before plunging into the Martian atmosphere, InSIGHT will deploy to cubesats to help track its progress during landing.

BELOW RIGHT: InSIGHT's 81-mile/130km-long landing ellipse in Elysium Planitia. The area is considered a top contender for the mission.

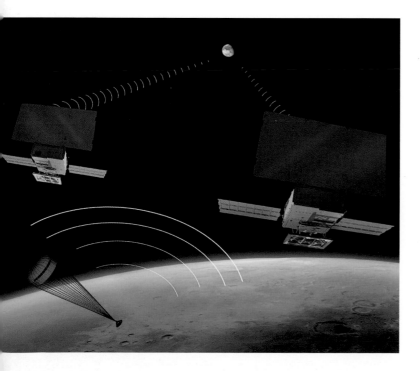

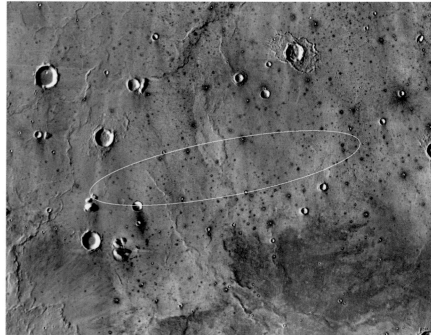

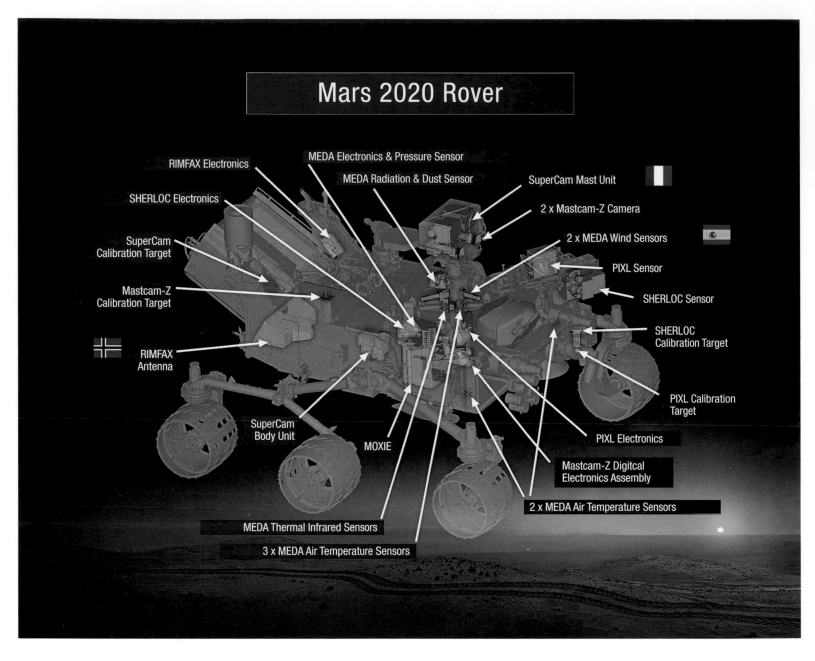

Mars 2020 Rover

RIMFAX Electronics

MEDA Electronics & Pressure Sensor

MEDA Radiation & Dust Sensor

SuperCam Mast Unit

SHERLOC Electronics

2 x Mastcam-Z Camera

SuperCam Calibration Target

2 x MEDA Wind Sensors

PIXL Sensor

Mastcam-Z Calibration Target

SHERLOC Sensor

SHERLOC Calibration Target

RIMFAX Antenna

PIXL Calibration Target

SuperCam Body Unit

PIXL Electronics

MOXIE

Mastcam-Z Digitcal Electronics Assembly

2 x MEDA Air Temperature Sensors

MEDA Thermal Infrared Sensors

3 x MEDA Air Temperature Sensors

capability. A more advanced instrument for determining organic compounds, the progenitor to life, will be included, as will a unit similar to Curiosity's ChemCam but with the ability to look for bio-signatures, or signs of life. There are other experiments (outlined below), but the real showstopper is the core-sample drill and sample-caching mechanism: the rover will take core samples and store them for future return to Earth for analysis, which could revolutionize our understanding of the Martian surface.

Specific instrumentation to be flown on the rover includes:

- MastCam-Z: A newly improved version of the cameras flown on Curiosity, but this time with the ability to zoom-in on objects

ABOVE: The Mars 2020 rover, set to be launched in 2020, is based on the successful Curiosity design. The internals will be far different, however, and include MOXIE, an experiment designed to test technologies to derive oxygen from the Martian atmosphere.

of interest (Curiosity had two sets of fixed lenses for wide and telephoto images).

- SuperCam: This is an upgraded version of Curiosity's ChemCam. Its abilities are similar—to analyze the mineral content of the remote targets by zapping them with a laser and reading the resulting flash. However, SuperCam will also be able to discern if organic substances exist within the surface of the rock or soil—a huge timesaver for investigating promising areas.

- The Planetary Instrument for X-ray Lithochemistry (PIXL): Like Curiosity's CheMin, PIXL will use X-ray fluorescence to determine the elemental composition of samples deposited into the instrument by the robotic arm, but in greater detail.

- Scanning Habitable Environments with Raman and Luminescence for Organics and Chemicals (SHERLOC): A spectrometer powered by an ultraviolet laser to determine organic compounds, whether living or pre-biological.

- Mars Environment Dynamics Analyzer (MEDA): A weather station that reports wind speed and direction, temperature, air pressure, humidity, and even the size and shape of airborne dust particles.

- Radar Imager for Mars' Subsurface Exploration (RIMFAX): A ground-penetrating radar unit that provides high-resolution information about the geological structure of the surface below the rover.

- Mars Oxygen ISRU (In-Situ Resource Utilization) Experiment (MOXIE): A device that will attempt to extract oxygen from the Martian atmosphere.

This final experiment, MOXIE, is intended as a precursor to human missions to Mars. If, as suspected, oxygen can be extracted from the CO_2-rich atmosphere of Mars, it can be used to generate breathable air, drinkable water, and rocket fuel production. The term ISRU stands for In-Situ Resource Utilization, and refers to the ability to use resources found on Mars (in situ, or on-site) to provide support for future crews on the planet.

The 2020 rover will carry a drill as Curiosity did, but rather than merely providing dust-like samples ground out of the rock and soil as Curiosity has done, this improved version will actually drill tubes into the surface to gather core samples. A total of thirty-one small tubes, each not much larger than a felt-tip pen, will be carried. As they are collected, the samples will be dropped-off in clusters for pick-up by a future sample return mission.

The 2020 rover will use the same sky crane landing system that successfully delivered Curiosity to Mars, with a few key changes. Landing sites are still being evaluated and the final choice is probably a couple of years off, but one top candidate is called Jezero Crater. At 28 miles (45km) wide, it is only about a quarter of the size of Gale Crater, Curiosity's base of operations. Landing will be tricky. However, after what has been found at Gale Crater by Curiosity, Jezero is appealing: it is another site where water definitely existed in great quantities for millions of years, and the geological diversity promises many rewards. Multiple erosion channels directed rocks and soil into a catchment basin,

or collection area, which should provide a vast assortment of targets for the sophisticated analytical instruments inside the rover. The long-lived body of water that existed there may even have once been a prime environment for Martian microbes, an added incentive for selection.

Other add-on technologies are being considered for inclusion on the mission. One compelling piece of technology is a small Mars helicopter—a drone-like unit that would depart from the rover, fly just high and far enough to image the terrain ahead, and transmit that data back for analysis. Modern rovers drive autonomously—with intermittent human oversight—and need the best information about the terrain ahead that they can get. Orbital maps are too imprecise, and the camera mounted atop the rover's mast has a limited range and ability to look over obstacles. A helicopter, or some other kind of airborne drone, could provide a vastly improved capability to see over and past rocks, sand dunes, and ridges.

There will likely be other changes to the mission—Mars 2020 is a work in progress, and final decisions will probably not be in place until 2018 or 2019, just in time for inclusion in the mission for the 2020 launch date. At the time of writing, there is no mission planned—or budgeted—to actually pick up and return back to Earth the core samples that the rover will collect.

The 2020 rover could provide a defining moment for a decision about sending humans to Mars. Assuming the MOXIE unit works as planned, confidence in our ability to collect oxygen from the Martian atmosphere would streamline mission planning substantially. And of course, if life or indicators of past life are found, it could change the way we view the red planet forever.

MARS 2020

MISSION TYPE: Mars rover
LAUNCH DATE: 2020 (projected)
LAUNCH VEHICLE: Atlas V (projected)
ARRIVAL DATE: 2020 (projected)
MISSION TERMINATION: N/A
MISSION DURATION: N/A
SPACECRAFT MASS: Approximately 2000lb (907kg)

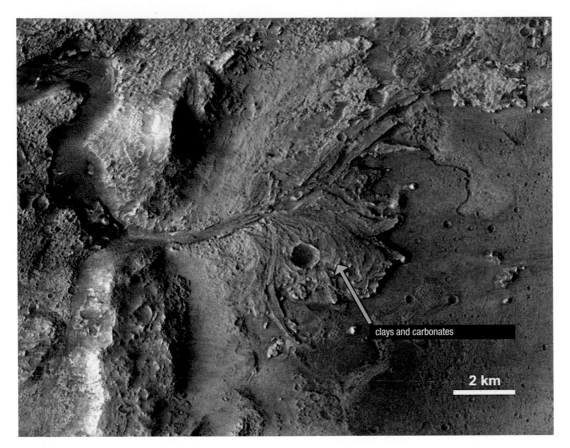

LEFT: Topping the shortlist of candidate landing sites for the Mars 2020 mission is Jezero Crater. The 28 mile/45km-wide formation is rich in water-formed geology.

BELOW: Jezero's lake had a long lifespan by Martian standards, and may have therefore once harbored life. The Mars 2020 mission has an astrobiology component, so Jezero has become an attractive choice.

OPPOSITE: If the Mars 2020 mission can determine that life once existed on the red planet, it will dramatically change the way we look at Mars, the universe and ourselves.

clays and carbonates

2 km

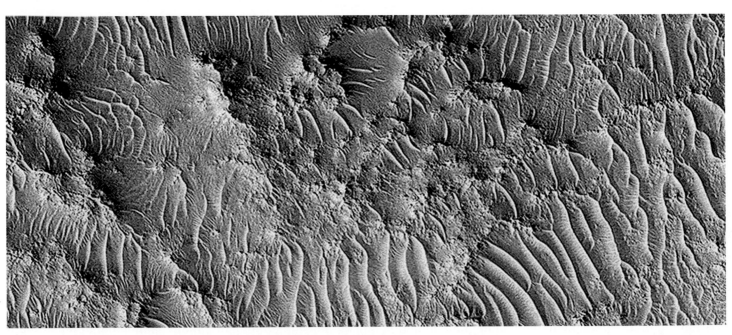

RUSSIA'S TURN: EUROPE'S EXOMARS

After the smashing success of Mars Express (*see* page 82), Europe started planning a new set of Mars missions. In 2008 and the next few succeeding years, ESA and NASA announced first one mission design, then another. The common ground seemed to be that the agencies would work in tandem, and there would be multiple spacecraft involved. First, an orbiter, similar in function to NASA's MAVEN (*see* page 150) would fly with a test lander.

THEN A ROVER ALONG WITH an associated lander would fly two years later, using the orbiter as a relay station. After ESA entered discussions to launch on a Russian Soyuz rocket, and NASA responded by offering a US Atlas V for the effort. The mission would fly sometime in 2016 or 2018. The plans flexed and changed, until in 2012, when NASA pulled the plug. The ugly specter of sequestration had crippled US science budgets, and cost overruns on other projects limited what was already committed. So the European consortium turned back to the Russians.

TGO AND SCHIAPARELLI

The twin ExoMars missions will be launched on two separate rockets, two years apart. The mission's name stands for Exobiology at Mars, and the search for life on Mars is its purpose. In a unique design approach, it consists of four separate spacecraft flying in two launches. In 2016, the Trace Gas Orbiter and the Schiaparelli lander headed off to Mars for arrival later that year. Then in 2018, a similar landing stage with a rover atop it will depart for the red planet. That's a total of one orbiter, two landers, and a rover.

Each component of the mission has specific goals. The 2016 Trace Gas Orbiter (TGO) will carry the Schiaparelli lander, and the two will part ways a few days before reaching Mars. The TGO

is designed to seek out small amounts of specific gasses in the Martian atmosphere, of which methane is the most compelling, due to its possible source as a result of biological activity. If propane or ethane are also found, either could bode well for a microbial origins. If sulfur dioxide is found, it would argue for a geological source of methane.

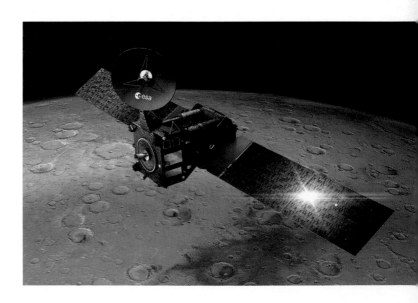

OPPOSITE: Upon nearing Mars, the 2016 Exomars TGO will eject the Schiaparelli lander. This Russian-designed component will attempt a safe touchdown on Mars, and is designed to return science data for up to a week.

ABOVE: The first component of the Exomars mission includes the Trace Gas Orbiter, launched in 2016. A primary goal is to search for methane in the Martian atmosphere.

The Mars Express orbiter spotted methane in the Martian atmosphere years ago, but it is an intermittent observation and the source has proved impossible to pin down. An indication of methane by TGO would be marvelous; if it is biologically originated, it will make history.

Instrumentation on the TGO includes one infrared and two ultraviolet spectrometers in an assembly named Nadir and Occultation for Mars Discovery (NOMAD). The Atmospheric Chemistry Suite (ACS), a Russian instrument, also uses an infrared spectrometer to analyze atmospheric makeup with extremely high precision. A neutron detector called the Fine Resolution Epithermal Neutron Detector (FREND), also a Russian contribution, will probe the first few feet of the Martian surface for hydrogen, and by extension, water ice. Finally, a camera package provided by the Swiss will provide high-resolution imaging.

The Schiaparelli Entry, Descent, and Landing Demonstration Module (Schiaparelli EDM) is a technology demonstrator, essentially a test-bed for the 2018 landing stage that will carry the ExoMars rover to the surface. The Schiaparelli EDM is about 8ft (2.4m) in diameter and shaped like a dish. It carries a small science package, but will only have power enough to operate for a few days, running off batteries. Both solar panels and a nuclear power generator had been discussed. Russia had considered providing a nuclear unit to ESA—but this never occurred and the lander was "de-scoped" or minimized to just a few days of battery-powered operation. After separating from the TGO, Schiaparelli will enter the Martian atmosphere and land using a heat-shield, parachute, and braking rockets. The final descent will be via rocket engines. When the lander makes contact with the surface, the remaining energy will be absorbed by a crushable stage below the main body of the lander (such a system was considered, but rejected, by the JPL teams for the Curiosity mission).

The landing site for Schiaparelli is Meridiani Planum, which is the same region in which the Mars Exploration Rover Spirit landed in (see page 92). Schiaparelli is equipped to make basic measurements of dust content of the atmosphere and weather measurements

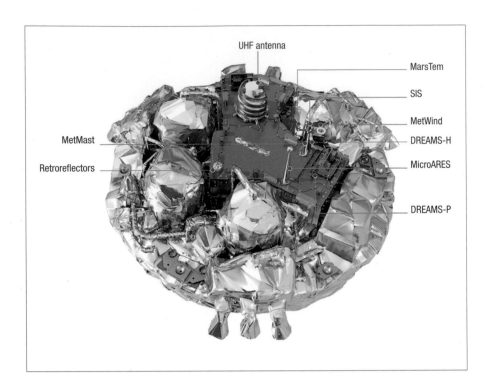

UHF antenna

MarsTem

SIS

MetWind

DREAMS-H

MetMast

MicroARES

Retroreflectors

DREAMS-P

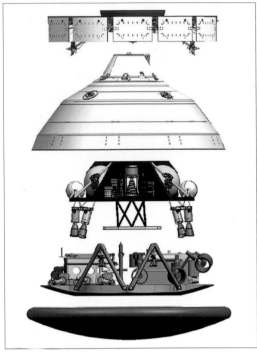

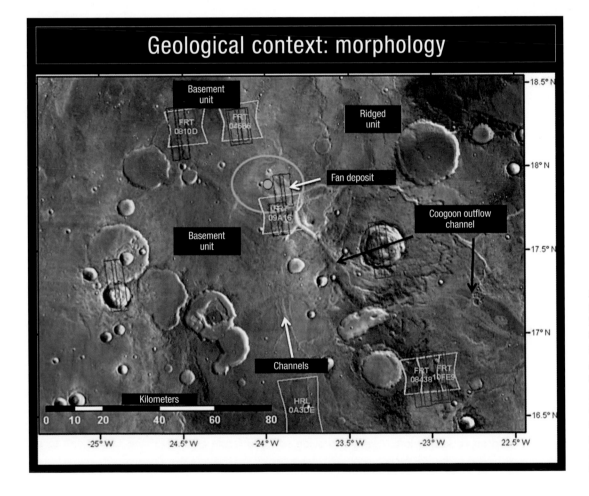

Geological context: morphology

18.5° N

Basement unit

FRT 0810D

FRT 04666

Ridged unit

18° N

Fan deposit

FRT 09A16

Coogoon outflow channel

Basement unit

17.5° N

17° N

Channels

FRT 08438

FRT 0FE5

16.5° N

Kilometers

0   10   20   40   60   80

HRL 0A3DE

-25° W   24.5° W   -24° W   23.5° W   -23° W   22.5° W

ABOVE LEFT: The Schiaparelli lander is primarily a test-bed for the later delivery of the Exomars 2018 rover, but does have scientific instrumentation and a battery to power them for a few days.

ABOVE RIGHT: The EDM configuration, showing the thermal protection shields.

LEFT: Oxia Planum contains one of the largest exposures of clay-bearing rocks and are around 3.9 billion years old.

OPPOSITE: A prototype of the ExoMars Rover at the 2015 Cambridge Science Festival.

(wind speed, temperature, air pressure, and surface temperature), as well as investigating the electrical fields on the surface, thought to be an important factor in the birth of Martian dust storms.

EXOMARS ROVER

It is hoped that TGO's orbital mission will last seven years after its arrival in October 2016. If all goes well with TGO and Schiaparelli, the second ExoMars launch is scheduled for 2018, though it may slip to 2020. This will use another Russian Proton rocket to loft a landing platform similar to the Schiaparelli lander, but with the

ExoMars 2018 rover strapped atop it. After touchdown, the 680-lb (308-kg) rover will depart the landing stage using ramps to drive down to the Martian surface.

The landing site for the ExoMars rover is currently Oxia Planum in the northern hemisphere of Mars, roughly halfway between the landing sites of Viking 1 and the MER Spirit rover (see pages 34 and 124). The region is composed largely of ancient clay beds dating back almost four billion years. It is hoped that the overlay of clays might provide soft and fertile ground for the rover's drill to gain access to deep, protected soil samples.

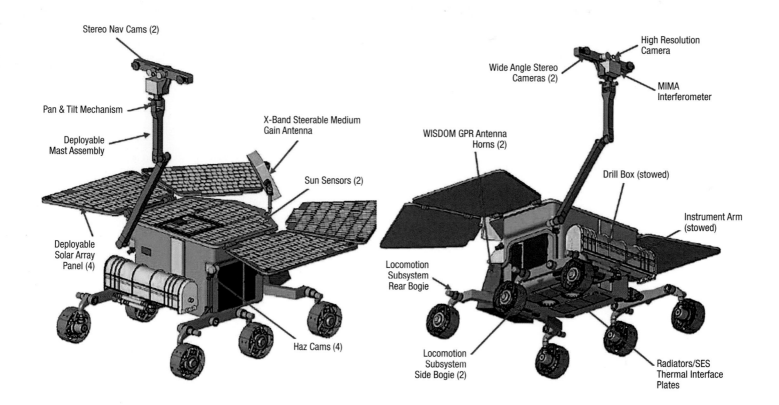

Stereo Nav Cams (2)

Pan & Tilt Mechanism

Deployable
Mast Assembly

Deployable
Solar Array
Panel (4)

X-Band Steerable Medium
Gain Antenna

Sun Sensors (2)

Haz Cams (4)

High Resolution
Camera

Wide Angle Stereo
Cameras (2)

MIMA
Interferometer

WISDOM GPR Antenna
Horns (2)

Drill Box (stowed)

Instrument Arm
(stowed)

Locomotion
Subsystem
Rear Bogie

Locomotion
Subsystem
Side Bogie (2)

Radiators/SES
Thermal Interface
Plates

Like the NASA Mars rovers, the ExoMars rover will be capable of driving autonomously. The TGO spacecraft will provide a radio relay to the Earth. The rover's primary goal is to find signs of life on the Martian surface. The instrument package it will use in pursuit of this goal includes: the Panoramic Camera system (PanCam), which will utilize wide-angle and telephoto lenses similar to those aboard Curiosity; an onboard life science lab, the Pasteur Instrument Suite, which includes the Mars Organic Molecule Analyzer (MOMA)—a highly evolved mass spectrometer—and an infrared imaging spectrometer (MicrOmega-IR), in conjunction with a Raman spectrometer (Ramen), which will analyze rock and soil samples ingested into the rover to search for organic molecules and biosignatures.

A ground-penetrating radar named WISDOM, capable of penetrating up to 10ft (3m) below the rover, will help identify promising sites for sampling. A robotic arm will contain another infrared spectrometer, the Mars Multispectral Imager for Subsurface Studies (Ma-MISS) inside the drill housing, and a close-up imager,

not unlike Curiosity's MAHLI, will also share space on the arm.

However, it may be the drill that is the star of this show. The rover will carry Mars' first core sample drill, capable of boring nearly 7ft (2.1m) below the surface in an attempt to find any micro-organisms that might be living beyond the reach of the intense radiation that bathes the planet. The Ma-MISS spectrometer will be able to image the boreholes from inside the drill. If all goes well, up to seventeen drilling attempts can be made.

There are many variables in the ExoMars plans. The first spacecraft, the Schiaparelli lander and TGO orbiter, were launched in early 2016. Assuming that the lander test item works well, there is still much development to do on the landing platform in order to deliver the rover to the surface in 2018. The Russian space agency is highly experienced in spaceflight and robotics, but has had negligible success to date with regard to Mars. With luck, 2018 will result in a successful partnership between ESA and Roscosmos in Russia for a fantastic mission to the red planet, including the possibility of the detection of extraterrestrial life.

NASA'S PLAN: GIVE US 25 YEARS

So—what is the next step after the robotic exploration of Mars? We have flown past the planet, orbited it, landed on it, and driven across its surface for over fifty years. We have analyzed its rocks and atmosphere, and will, in the not too distant future, likely bring home samples of the planet for direct evaluation here on Earth. For many, it feels as though the time has come to send *people* there…

OF COURSE, THERE ARE MANY WHO WILL ASK, why go? This same question came up during the Apollo moon program in the 1960s. "Why not spend the money solving Earth's problems instead of running off to explore another place?" many asked. There are a few answers, and two of them take prominence.

First of all is the science. NASA engaged a group of scientists and charged them with the task of specifying some of the scientific rationales for a crewed Mars project, and they outlined the advantages of traveling to the planet clearly. In sum, a human on Mars with the proper equipment could accomplish, in just one day, many of the tasks it takes a rover as much as a decade or more to do. Overland travel can be accomplished much more rapidly with a human at the controls—the MER rover Opportunity took ten years to cross 26 miles (42km) of terrain, a distance an astronaut in a rover could do quickly. The Apollo 17 astronauts, using their lunar rover, covered 22 miles (35km) on the moon in just three days with 1960s technology. Humans are also able to rapidly interpret new data—using what they see and measure to formulate changes in the science plan—and adapt immediately. Human dexterity is well beyond anything robotics will be able to accomplish for a decade or more. There are many other reasons, but these are some of the key ones.

Then there is the question of survival. Earth is subject to many maladies—runaway climate change, debilitating pandemics, global warfare and impacts from cosmic bodies like asteroids. Any of these could greatly diminish our chances for survival. Having Mars as a second outpost for humanity greatly increases our chances

RIGHT: The brilliant German rocketeer Wernher von Braun first published his ideas about voyaging to Mars in *Das Marsprojekt*, first published in 1949.

BELOW: An early vision of Mars exploration by NASA.

OPPOSITE: Walt Disney, left, poses with Wernher von Braun, right, in 1954. Von Braun, then working for the US Army, was engaged by Disney to advise on a series of television shows about the future of space exploration.

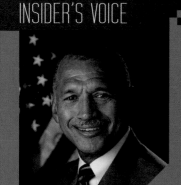

Charles Bolden

NASA Administrator

Charlie Bolden is an enthusiastic promoter of NASA's overall plan to send humans to Mars. One part of this plan is sending supplies and machinery ahead of the astronauts, and conventional rockets take too long and cost too much. He likes the alternative: solar-electric rockets.

"We are working very hard to help us get to the point where we can get astronauts to Mars much quicker than today. Right now it's about an eight-month mission; we'd like to cut that in half if we can.

"We're looking at how we get cargo and supplies to Mars, because we have to take a lot of stuff when we talk about putting humans on the Martian surface. So we are looking at solar electric propulsion systems … what we are trying to do now is increase the capability of those systems—whether we might cluster multiple engines together. The limiting factor of that type of propulsion is electric power to drive it, so [we are] trying to get more energy density onto a solar cell. The more power we can get, the larger we can make the engine and its capability.

"We are looking for that kind of propulsion to get cargo way out in front of the astronauts—they can catch up with it and pass it on the way to Mars!"

of continuing as a species should we be faced with a major environmental challenge on Earth. As the great physicist Stephen Hawking has said, "The human race has no future if we do not go into space," and Mars is the best option for off-Earth human settlement in the solar system.

As we have seen, the question of exploring Mars has been under discussion since at least the late 1940s, when Wernher von Braun's groundbreaking book *The Mars Project* laid out an ambitious plan that might have worked, had Mars been what we then thought it was. However, his mission design was on a scale that would have made the Manhattan Project to develop the first atomic bomb look inexpensive in comparison. In any case, once Mariner 4 reached Mars in 1965, we knew that von Braun's plan was unworkable—the planet is a tougher place to visit than was previously thought.

As more probes visited the planet, the news just got worse… Mars has a thin, unbreathable atmosphere that would require human beings to wear pressure suits in order to survive exposure to it. The temperatures there are unbearably cold. Astronaut-roasting radiation bathes the surface of Mars continually, and its soil is laced with perchlorate—a substance toxic to humans, even in low concentrations. The discoveries of the past five decades have not been kind to plans for human Mars exploration. But still, the goal has persisted.

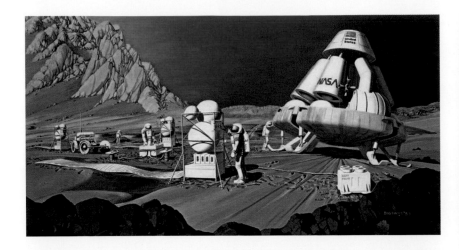

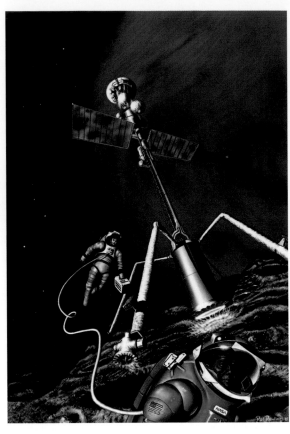

PUTTING LIFE ON MARS

As the Apollo lunar landing effort continued throughout the 1960s, NASA considered various follow-on programs that might continue their incredible thrust into deep space. Long-term moon bases were one option. An Earth-orbiting space station was another. And, of course, the trek to Mars was an ongoing consideration. In the end, we got Skylab—a small and temporary orbital outpost—and the space shuttle. A crewed Mars mission was off the agenda. The US government did not feel that there was enough popular support for such a vast technological undertaking after the expensive rigors of Apollo. At the same time, the Soviet Union was continuing to attempt the robotic exploration of Mars, with little success. Plans for a Soviet human exploration program came and went, but with their moon-landing program resulting in expensive failure, it was all the USSR could do to fly their own space stations in Earth orbit.

Nonetheless, plans for a crewed mission to Mars continued. In the 1970s and '80s, the various aerospace contractors that had helped America reach the moon turned out study after study of how a manned Mars mission might be flown. Some ideas were more feasible than others but, ultimately, none got past the starting line—there was no money and little support outside the space agency. NASA then began a series of Mars mission design studies to spell out what kind of technology would be required to undertake such an endeavor, should it one day be approved and funded. These were called the Mars Design Reference studies, with the first being completed in 1993 and the fifth iteration arriving in

2009. In general terms, each version was a bit more streamlined and technologically sophisticated than its predecessor.

The usual approach of these plans has been to limit exposure to radiation both in space and on Mars, to look at both traditional chemical and alternative propulsion systems (such as solar-electric and nuclear), and to explore ideas about sending cargo ahead of the astronauts. Additionally, ways to develop in situ or on-site resources have been considered, as well as various options in respect of the exploration of the planet once astronauts have arrived there. Other studies have examined the possibility of crewed landings on Martian moons, which would be much easier to achieve than exploring the planet itself.

When the space shuttle program was concluded in 2011, NASA had no flyable replacement for it. It did, however, have an International Space Station the size of an American football field passing 200 miles (322km) overhead every ninety minutes. To maintain access, NASA now purchases rides to the ISS from the Russian space agency Roscosmos in their Soyuz capsules (which, ironically, were an evolution of the same spacecraft the Soviet Union had built to beat America to the moon). Meanwhile, NASA slowly continues work on the Orion capsule, and SpaceX and Boeing have both procured contracts to deliver both astronauts and cargo to the space station for NASA. All this is a work in progress as of 2016 but, from a Mars-centered perspective, NASA has proclaimed that the Orion capsule is being designed with a Mars mission firmly in mind. Their plan is called the Flexible Path, and the crewed Mars portion is referred to as the Evolvable Mars Campaign. Both are attempts to grow the

OPPOSITE TOP LEFT: An artist's concept of a NASA Mars mission, circa 1989.

OPPOSITE BOTTOM LEFT: A contemporary artist's impression of a Mars expedition from the most recent Mars Design Architecture study. Note the inflatable habitat atop the lander in the background.

OPPOSITE RIGHT: A 1986 artistic rendering of a Mars outpost on Phobos.

ABOVE: The massive Space Launch System is NASA's current project to accomplish deep space exploration, the primary goal of which is a journey to Mars. The Orion spacecraft rides at the top. The side boosters are enlarged versions of the space shuttle's solid rocket boosters, and the four engines clustered at the base are recycled shuttle main engines.

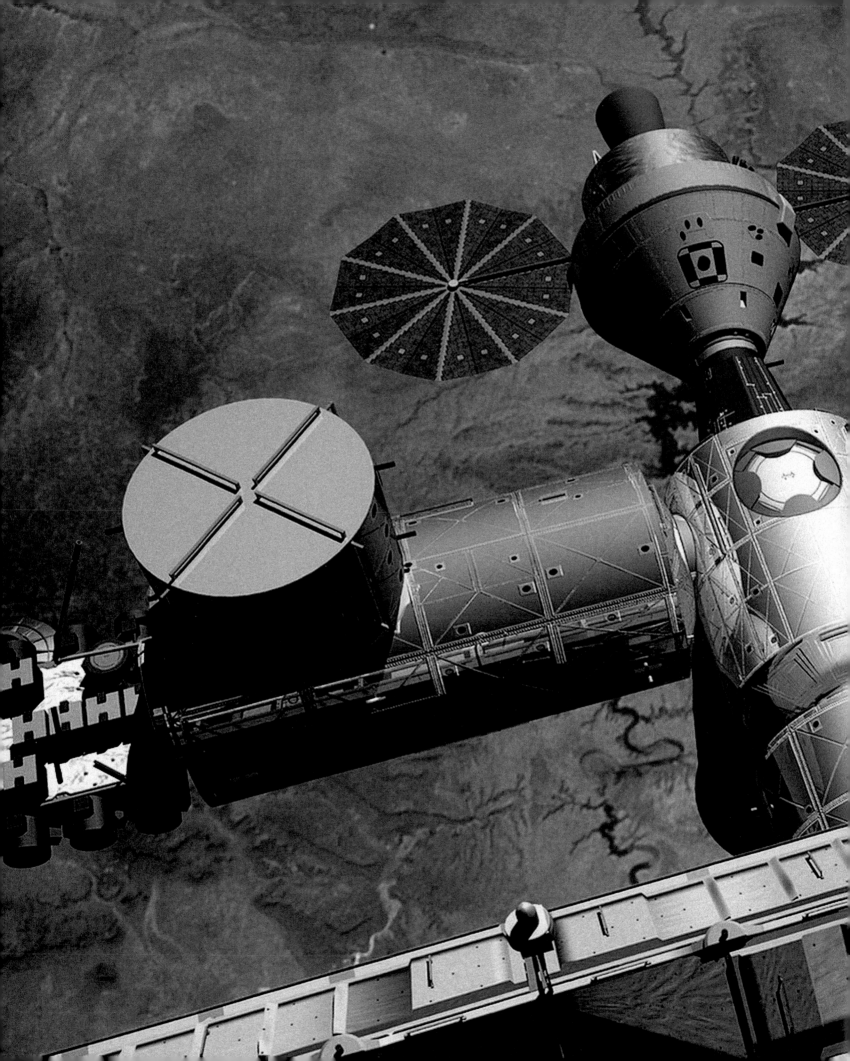

agency's goals for a human mission to Mars within the realities of ever-shifting federal budgets and uncertain financial support.

THE FLEXIBLE PATH

The origins of the Flexible Path can be traced to the 2009 cancellation of the Constellation Mission, which had been initiated under President George W. Bush in 2004. Rather than Constellation's structured plan for returning astronauts to the moon, the Flexible Path advocates a number of options for the new hardware currently under construction—the Orion spacecraft and the Space Launch System mega-booster. These include:

- Lunar flybys and possible near-lunar crewed stations
- Deep-space stations at the Lagrange points (stable orbital points between the Earth and the moon, and beyond)
- Crewed Mars or Venus flybys
- Landings on Mars or its moons
- Missions to asteroids

The last of these, the asteroid mission, became the primary goal of NASA's crewed efforts beyond the space station for the 2010–30 timespan. Initially designed to send a crew to an asteroid deep in space, this plan was later modified to bring a small piece of asteroid closer to the moon for a visit by humans. The core of the plan, called the Asteroid Redirect Mission, was that the technologies needed to accomplish this would help evolve needed expertise for long-term Mars mission planning.

While the logic may be accurate to a degree, it has fallen flat. Those old enough to remember saw NASA fly to the moon every two months in the 1970s, build two space stations—one in the 1970s and another in the 1990s—and send rovers to Mars to explore the surface for a decade at a time. After these accomplishments, the ideas of exploring a 10ft (3m) chunk of relocated asteroid has not been as popular as the agency had hoped. Poll after poll declares that, if we are to pursue human spaceflight at all, the moon and Mars are hands-down favorites. Notably, many of NASA's top astronauts from the Apollo years and beyond agree on this point.

NASA's current plan dictates no crewed spacecraft flying anywhere near Mars before 2035, and possibly later. So, the agency's long-term plans involve landing on Mars (or one of its moons) in the late 2030s. It has been challenging to gain sufficient funding from Congress and the Executive Branch to pay for the mission. We are still supporting an expensive space station and dozens of robotic programs. While NASA has planned to reach Mars within its current budget projections, it would be far easier—and faster—to work with larger funding levels. But an Apollo-level commitment—which was, at its height, almost five percent of the federal budget—has simply not materialized for Mars (or any other

destination). NASA's budget is currently a tenth of that figure, and developing the needed hardware within these constraints will be difficult. Given all this, what is most likely to occur?

For NASA, there will continue to be slow, incremental development across a broad spectrum of technology aimed at the ultimate goal of humans reaching Mars. The recent one-year mission of astronaut Scott Kelly to the space station provided much-needed data on the long-term effects of weightlessness on the human body. Data returning from the continuing mission of Curiosity and the upcoming 2020 mission will illuminate the ongoing risks from radiation, and provide experimental results of in-situ resource utilization. NASA will continue working towards crewed testing sometime in 2021–23, and the SLS should launch its first test flight in 2018.

Another part of the crewed Mars infrastructure is an inflatable space habitat built by a company named Bigelow Aerospace. A prototype similar to what may be used for the transit to Mars is scheduled to be tested on the space station in the next few years. This would add many hundreds of cubic feet of available living space to the 700ft^3 (213m^3) Orion spacecraft as it heads for Mars.

Of course, other nations are aiming for Mars as well, but none has, at this point in time, NASA's experience in deep-space flight or landing on another world. Russia regularly announces its human Mars ambitions, and a few years ago completed a 500-day Earth-based simulation of such a flight, but little beyond this is yet formally funded. Europe's ESA has suggested crewed Mars missions, both in co-operation with Russia or alone, but recently refocused their human spaceflight efforts on the moon. China has flown crewed spacecraft since 2003, and even orbited a small space station, but their near-term goals are directed toward a larger orbital station and landing humans on the moon. They have said that a crewed Mars mission is a possibility sometime between 2040 and 2060.

At this point, NASA appears to be the only national space agency with serious ambitious for the Martian journey. However, a 2014 report from the National Research Council warns that if NASA is serious about such an undertaking, a number of criteria must be met. These are: budgets must be increased; Mars-directed efforts prioritized; unnecessary and non-related programs discontinued; and international participation encouraged. Finally, and most importantly, an overall Mars-centered plan must be approved and implemented, without the changes and cancellations that so often occur when new presidential administrations come into office.

These recommendations are sound, but two years later have still not been implemented. NASA continues to develop components of a Mars program that should eventually result in astronauts exploring and living on the Martian surface. Only time will tell how quickly these plans are realized.

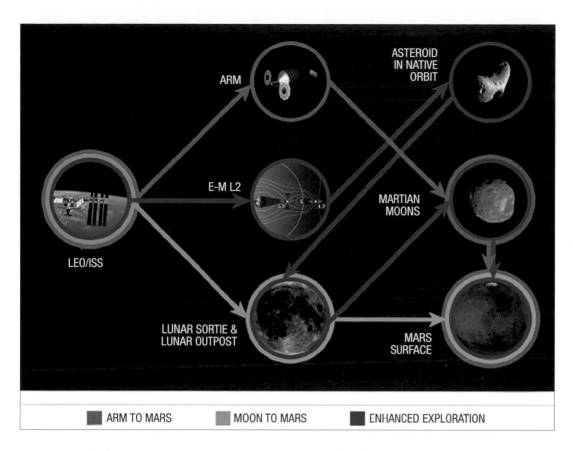

PREVIOUS PAGE: An artist's impression of NASA's current Mars plan: the Orion capsule, top left, is docked with a Mars orbiting complex based largely on International Space Station technology.

LEFT: NASA's Mars plan as of 2016. From left: spacecraft depart Low Earth Orbit (LEO) to three possible destinations—asteroid rendezvous near the moon (ARM), E-M L2 (Lagrange point 2 near the moon), or a lunar outpost. From there, it is off to Mars, either one of the planet's moons or the surface. Top right is a rendezvous with an asteroid in deep space.

BELOW: An artist's impression of a Mars base with a food production module, cutaway to left. While it is believed that some plants can thrive in Martian soil (including asparagus), it is likely that most food production will occur in a controlled environment.

ARM

ASTEROID IN NATIVE ORBIT

E-M L2

MARTIAN MOONS

LEO/ISS

LUNAR SORTIE & LUNAR OUTPOST

MARS SURFACE

■ ARM TO MARS   ■ MOON TO MARS   ■ ENHANCED EXPLORATION

MARS ROUTE ONE

NASA's grand plan is one way that humans could be sent to Mars, but there are alternatives to their approach. Some of these plans are larger and would be far more expensive to undertake, while others attempt to trim costs and accelerate the schedule. Given the uncertain nature of Mars funding for NASA over the next twenty years, a few of these well-engineered alternative proposals deserve consideration.

AS PREVIOUSLY DISCUSSED, A HUMAN MISSION TO MARS has been on the minds of engineers, scientists, writers, futurists, and many others for decades. After the moon, Mars has long been the primary human goal in the solar system. Brilliant minds from dozens of countries, multiple space agencies, and hundreds of academic institutions have created workable plans to support this endeavor. These mission designs come in many shapes and sizes, ranging from fast, simple crewed flybys of the planet to multiple landings with outposts and even colonies.

THE ALDRIN CYCLER

Buzz Aldrin, renowned Apollo astronaut and second man to walk on the moon, has ideas about humanity's trek to Mars. In fact, he has written entire books about the subject of the future of America's space program, the most recent entitled *Mission to Mars*. In Aldrin's view, we should take pause and regroup.

Aldrin's plan—and it's a large one—unites international space agencies and commercial ventures in an interlocking program to move beyond Earth orbit and establish a presence on the moon prior to traveling to Mars. The moon, and the space surrounding it, are good places to set up preliminary outposts that can be used for everything from research and training to hardware manufacture and assembly before traveling to Mars. Aldrin also points out that there appear to be large volumes of water at the lunar poles, which can be used to create fuel and other consumables in space—so no need to launch that from Earth. The attendant cost savings would be enormous.

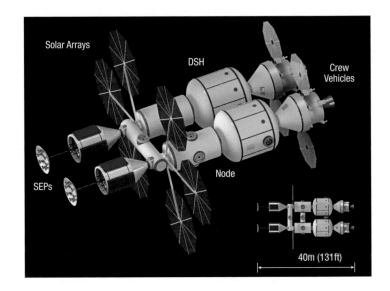

Aldrin's plan urges NASA to solicit the co-operation of international powers and private entrepreneurs. NASA benefits from cost reduction, and the other entities benefit from NASA's vast base of knowledge, gained from decades of going to the moon and working in space. With this international network of outposts and research stations, we can better prepare for a sustained effort not simply to explore Mars, but to develop a sustainable infrastructure in which to make humanity a multi-planet species.

The second part of this vision is known as the "Aldrin Cycler." This is a long-term logistics program that allows for repeated crew

---

ABOVE: The Aldrin cycler would complete a one-way trip to mars in five months, and continue to loop between Mars and Earth permanently.

OPPOSITE: By adding modules to the Orion capsule for the Mars-bound trip—propulsion, habitation, cargo storage, and more—comfort and safety are enhanced.

and cargo trips to Mars without much of the expense of launching each journey afresh. In sum, the cycler is a larger Mars craft that would be assembled in space and sent into a long, looping, and permanent orbit between Earth and Mars. Once there, the energy (and fuel) needed to keep the cycler moving between the two planets would be minimal. Then, whenever a trip needs to be made to or from Mars, the crew need only to be ferried to or from the cycler. It is a bit like taking a taxi to the airport and flying from New York to Los Angeles on a United Airlines flight, instead of making the entire drive yourself.

With a couple of cyclers in operation and the proper infrastructure on each end—ferries, outposts, and fuel depots—moving people and cargo to Mars and back would be greatly simplified and could become a regular occurrence. However, it would be an expensive program to mount, and would require a large commitment up front, which is why Aldrin supports the idea of international entities working together, along with the private sector, to make it happen. Such a program could save very big money in the long run.

MARS DIRECT

Another well-respected plan was put forth by an aerospace engineer named Robert Zubrin. He is the co-founder of the Mars Society and has a PhD in nuclear engineering, as well as multiple masters' degrees in nuclear and aeronautical engineering. So, when Zubrin wrote a series of books, starting with *The Case for Mars*, published in 1996, people paid attention. Zubrin described a mission architecture he called Mars Direct, which was an attempt to streamline NASA's larger and more complex designs for Mars voyages. Zubrin's contention was that this process could be simplified by sending robotic spacecraft ahead of the astronauts, which could begin to extract useful materials from the Martian environment. Astronauts would then launch directly from Earth to the red planet. When they arrived, the materials they needed to survive would already be at hand, and far less mass would would have to be carried aloft, so the mission parameters could be dramatically shrunk.

This is not to say that NASA, Aldrin and others don't intend to use local resources—most Mars mission designs now do—it is simply that Zubrin and his collaborators came early to the idea.

BOEING'S PLAN

The Boeing Corporation issued a plan in 2014 that is based on full utilization of our existing space architecture, including the International Space Station and the US' new Space Launch System (SLS) rocket.

The key elements for this mission include NASA's Orion spacecraft, a "space tug" powered by an electric engine that runs off solar power, an inflatable habitat module for the crew while in space, and a Mars lander and ascent vehicle.

The journey begins with an SLS launch of the space tug, and a robotic (un-crewed) cargo-carrying lander. A second un-crewed launch sends the habitat and return rocket stage to rendezvous near the moon with the cargo vessel and tug. These four modules can depart for Mars when they are ready and a launch opportunity occurs, and then follow a slow trajectory to Mars, powered by the electric rocket engines. Once it arrives after about five hundred days, the cargo lander descends to the Martian surface.

When these elements are safely delivered to Mars, two more SLS launches carry another habitat module with a solar-electric stage, the Orion spacecraft, and a Mars lander and ascent vehicle to rendezvous near the moon. This assembly then departs for Mars, following a faster trajectory to cut transit time to about 256 days.

Once the crew arrives in Martian orbit, it transfers to the lander and descends to Mars, landing near the previously emplaced habitat. After the stay, when the time is right for a proper return trajectory, the crew launches and conducts a rendezvous with the orbiting complex in order to return home.

It sounds complicated, but Boeing's mission design is one way in which to maximize the potential of the spaceflight architecture that NASA is developing now, and utilizes a reasonable rate of launches for the SLS. A key element of this approach is to try and stay within NASA's current budgets levels and avoid an Apollo-style crash-program of space technology development. The plan includes contingencies for precursor missions to asteroids and the Martian moons for technology development and probing, before the big journey to the Martian surface. The first humans to reach Mars may even explore it from the moon Deimos, remotely operating robotic units on the surface.

SPACEX

The wildcard in this discussion is Elon Musk and SpaceX. Since founding the company in 2002, Musk has maintained that a primary goal is the colonization of Mars. How this would be accomplished and what the profit model (if any) might be is less clear, but given the company's impressive achievements in the Earth-orbital arena—in a decade, revolutionizing the commercial launch services business and creating domestic competition where little existed—it's a goal worth noting.

Plans include the design and construction of a Mars Colonial Transporter (MCT), an evolution of SpaceX's current Falcon 9 and future Falcon Heavy rocket. The first stage of the MCT would be roughly equivalent to three 33ft/10m-diameter Apollo-era Saturn V rockets strapped together, though a single, 50ft/15m-diameter mega-booster has also been discussed. The specifics remain to be announced, but it is hoped that the rocket would be ready to launch by the mid-2020s with an ultimate goal of sending 100 tons of payload to Mars. It's an impressive target. Musk's expressed hope is to have tens of thousands of people living and working on Mars by the mid-twenty-first century.

Musk has said a number of times that he feels that it's imperative for humanity to establish a foothold on another world to assure its survival, and that Mars is the best option. Regardless of financial returns, he feels that the risks of being dependent on a single planet are just too high.

INTERNATIONAL EFFORTS

Europe's ESA and Russia have suggested their own intentions to travel to Mars, either individually or as partners. The general concepts under consideration are similar to those of the US, with the Russian plan looking in many respects like a blend of NASA and Boeing's plans. The Russians are currently building a replacement spacecraft for the venerable Soyuz, and have begun designing their own replacement for the aging International Space Station, due for retirement in 2024. They have discussed various routes to Mars from there, including some that start from points out by the moon. Nuclear or solar-electric spacecraft are under consideration, with a Mars orbital "base" that would dispatch and receive landers and ascent vehicles. There has been talk of joint missions—and, after all, the US and Russia teamed up to build the ISS—but a difficult history in space partnerships and the current geopolitical situation make this unlikely to happen in the near future. It should be noted that Russia (as the Soviet Union) has vast amounts of experience in spaceflight, especially with their space stations MIR and Salyut. The Russians have also spearheaded five-hundred-day simulations of Mars journeys on the ground. While they have never sent cosmonauts beyond the magnetosphere, they are clearly capable of much in space.

LEFT: SpaceX's newest capsule, the Dragon 2, may one day be outfitted as a Mars landing vehicle.

OPPOSITE: The European Space Agency has its own Mars plans, possibly in collaboration with Russia. For now, however, they are looking closely at establishing a lunar base prior to heading off to the red planet.

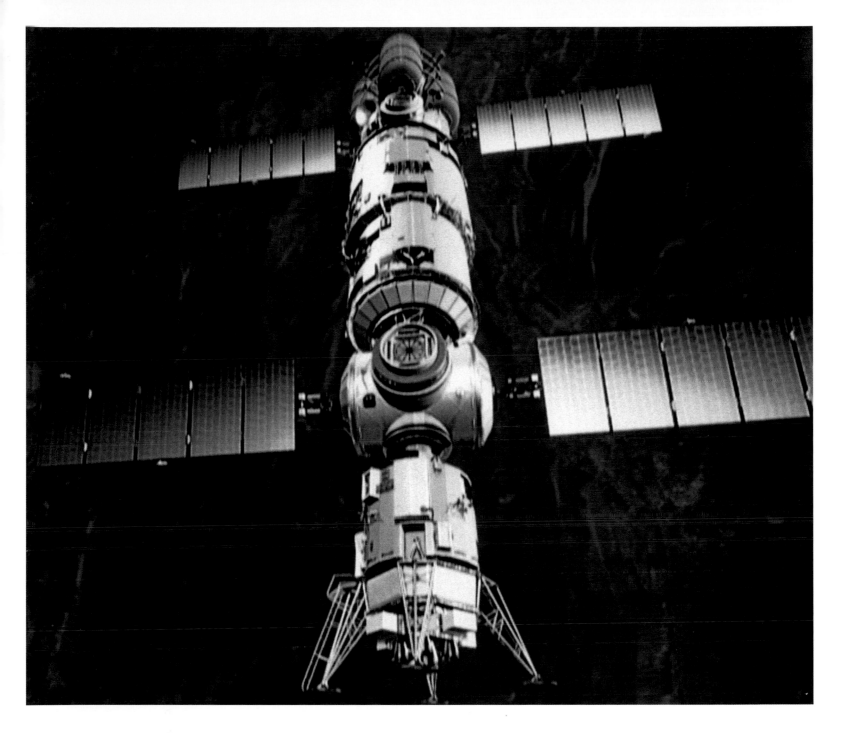

VAST CHALLENGES

To date, only the US has experience with human crews beyond Earth orbit. However, the challenges are exactly the same for any nation planning the voyage. These include:

- Building/assembling/manufacturing large Mars-capable spacecraft in Earth orbit or beyond.
- Crew support—food, water, and a breathable air supply—during the trip to Mars and back.
- Radiation protection, both during the journey and while on the surface of Mars.

- Landing heavy spacecraft on Mars with its thin atmosphere (crewed spacecraft will be many times the mass of the robotic landers that have arrived safely to date).
- Carrying (or extracting) enough fuel to return home.

And there is of course the question of money. Any venture to the red planet will be expensive and require engineering and technology currently unavailable, multiple launches of large rockets, and the infrastructure to build and oversee the missions from the ground. Mars will not allow humans to visit its surface without a struggle. Using resources already found in space, and on the Martian surface, will greatly reduce the requirements for lifting heavy supplies into orbit and sending them off to Mars. While a crewed Mars mission could be accomplished without this in situ resource utilization, it would be staggeringly expensive. This is one of the reasons why the Mars 2020 rover will include the MOXIE experiment to pioneer and test the extraction of oxygen from the Martian environment.

One way to simplify sending humans to the surface of the red planet is simply to leave them there once they arrive—the one-way trip option. A number of private pro-Mars exploration advocacy groups have expressed conditional support for such plans. The reduction in mass, and therefore launches from Earth, make this an attractive course to follow if the survival of the crew can be assured over the long-term. Studies of this type of mission arrive at mixed conclusions. Some see the one-way-to-stay option as achievable with moderate risk, while others see many more challenges. In any case, more research must be performed before anyone can be sent on a one-way ride to Mars, and most agree an option to return would eventually need to be provided.

A robotic mission to return Martian soil samples, rocks, and atmospheric samples to Earth for analysis and experimentation could make the mission planning process much more precise, which is part of the reason why the Mars 2020 rover will sample the surface and "cache" these samples for eventual return to Earth. If these samples can be taken into a laboratory, either on Earth or in orbit, we will have a much better idea of how to capitalize on their resources. We have a reasonable idea of the mineral composition of many parts of Mars already, but having the rocks and soil in hand would allow for direct testing of the materials for maximum utilization and safety.

Humanity's journey to Mars has been a long one. It began in the minds of the ancients, as they created powerful beings to inhabit the place of the red planet in our collective psyche. Mars then became a place, a neighboring world, thought to be a sister planet to Earth. In the last century, Mars was revealed as a cold, windswept, and rocky planet, devoid of any apparent life and with only a thin and tenuous atmosphere. Only in the past two decades has the more moderate truth about Mars finally come to light: it is still a hostile planet, but it seems to have the elements necessary for supporting life—either microbial in the past, or human in the future. It will simply take more robotic investigation, and ultimately a journey by human beings, to address our endless curiosity about the red planet, and to enable us to call it home.

RIGHT: NASA has created a number of plans for human Mars exploration. This image dates from a recent 2006 effort. Two astronauts inspect a robotic lander and rover as their own vehicle awaits. The landing site is seen in the background left.

MARS EXPLORATION TIMELINE

Marsnik 1/Mars 1960A

480 kg—USSR Mars Probe—(October 10, 1960)
Failed to reach Earth orbit.

Marsnik 2/Mars 1960B

480 kg—USSR Mars Probe—(October 14, 1960)
Failed to reach Earth orbit.

Sputnik 22 (Mars 1962A)

USSR Mars Flyby—900 kg—(October 24, 1962)
Spacecraft failed to leave Earth orbit after the final rocket stage exploded.

Mars 1

USSR Mars Flyby—893 kg—(November 1, 1962)
Communications failed en route.

Sputnik 24 (Mars 1962B)

USSR Mars lander—mass unknown (November 4, 1962)
Failed to leave Earth orbit.

Mariner 3

USA Mars Flyby—260 kg—(November 5, 1964)
Mars flyby attempt. Solar panels did not open, preventing flyby. Mariner 3 is now in a solar orbit.

Mariner 4

USA Mars Flyby—260 kg—(November 28, 1964–December 20, 1967)
Mariner 4 arrived at Mars on July 14, 1965 and passed within 6118 miles (9846km) of the planet's surface after an eight-month journey. This mission provided the first close-up images of the red planet. It returned 22 close-up photos showing a cratered surface. The thin atmosphere was confirmed to be composed of carbon dioxide in the range of 5–10 mbar. A small intrinsic magnetic field was detected. Mariner 4 is now in a solar orbit.

Zond 2

USSR Mars Flyby—996 kg—(November 30, 1964)
Contact was lost en route.

Mariner 6

USA Mars Flyby—412 kg—(February 24, 1969)
Mariner 6 arrived at Mars on February 24, 1969, and passed within 2136 miles (3437km) of the planet's equatorial region. Mariner 6 and 7 took measurements of the surface and atmospheric temperature, surface molecular composition, and pressure of the atmosphere. In addition, over 200 pictures were taken. Mariner 6 is now in a solar orbit.

Mariner 7

USA Mars Flyby—412 kg—(March 27, 1969)
Mariner 7 arrived at Mars on August 5, 1969, and passed within 2206 miles (3551km) of the planet's south pole region. Mariner 6 and 7 took measurements of the surface and atmospheric temperature, surface molecular composition, and pressure of the atmosphere. In addition, over 200 pictures were taken. Mariner 7 is now in a solar orbit.

Mars 1969A

USSR
Launch Failure

Mars 1969B

USSR
Launch Failure

Mariner 8

USA Mars Flyby—997.9 kg—(May 8, 1971)
Failed to reach Earth orbit.

Kosmos 419

USSR Mars Probe—4549 kg—(May 10, 1971)
Failed to leave Earth orbit.

Mars 2

USSR Mars Orbiter/Soft Lander—4650 kg—(May 19, 1971)
The Mars 2 lander was released from the orbiter on November 27, 1971. It crashed-landed because its braking rockets failed—no data was returned. This was the first human artifact to reach the surface of Mars. The orbiter returned data until 1972.

Mars 3

USSR Mars Orbiter/Soft Lander—4643 kg—(May 28, 1971)
Mars 3 arrived at Mars on December 2, 1971. The lander was released and became the first successful landing on Mars. It failed after relaying 20 seconds of video data to the orbiter. The Mars 3 orbiter returned data until August, 1972. It made measurements of surface temperature and atmospheric composition.

Mariner 9

USA Mars Orbiter—974 kg—(May 30, 1971–1972)
Mariner 9 arrived at Mars on November 3, 1971 and was placed into orbit on November 24. This was the first US spacecraft to enter an orbit around a planet

other than Earth. At the time of its arrival a huge dust storm was in progress on the planet. Many of the scientific experiments were delayed until the storm had subsided. The first hi-resolution images of the moons Phobos and Deimos were taken. River and channel like features were discovered. Mariner 9 is still in Martian orbit.

Mars 4

USSR Mars Orbiter—4650 kg—(July 21, 1973)
Mars 4 arrived at Mars on February, 1974, but failed to go into orbit due to a malfunction of its breaking engine. It flew past the planet within 1379 miles (2200km) of the surface. It returned some images and data.

Mars 5

USSR Mars Orbiter—4650 kg—(July 25, 1973)
Mars 5 entered into orbit around Mars on February 12, 1974. It acquired imaging data for the Mars 6 and 7 missions.

Mars 6

USSR Mars Orbiter/Soft Lander—4650 kg—(August 5, 1973)
On March 12, 1974, Mars 6 entered into orbit and launched its lander. The lander returned atmospheric descent data, but failed on its way down.

Mars 7

USSR Mars Orbiter/Soft Lander—4650 kg—(August 9, 1973)
On March 6, 1974, Mars 7 failed to go into orbit about Mars and the lander missed the planet. Carrier and lander are now in a solar orbit.

Viking 1

USA Mars Orbiter/Lander—3527 kg including fuel—(August 20, 1975—August 7, 1980)

Viking 2

USA Mars Orbiter/Lander—3527 kg including fuel—(September 9, 1975—July 25, 1978)
Viking 1 and 2 spacecraft included orbiters (designed after the Mariner 8 and 9 orbiters) and landers. The orbiter weighed 883 kg and the lander 572 kg. Viking 1 was launched from the Kennedy Space Center, on August 20, 1975, for the trip to Mars and went into orbit about the planet on June 19, 1976. The lander touched down on July 20, 1976 on the western slopes of *Chryse Planitia* (Golden Plains). Viking 2 was launched for Mars on November 9, 1975, and landed on September 3, 1976. Both landers had experiments to search for Martian microorganisms. The results of these experiments are still being debated. The landers provided detailed color panoramic views of the Martian terrain. They also monitored the Martian weather. The orbiters mapped the planet's surface, acquiring over 52,000 images. The Viking project's primary mission ended on November 15, 1976, eleven days before Mars' superior conjunction (its passage behind the Sun), although the Viking spacecraft continued to operate for six years after first reaching Mars. The Viking 1 orbiter was deactivated on August 7, 1980, when it ran out of altitude-control propellant. Viking 1 lander was accidentally shut down on November 13, 1982, and communication was never regained. Its last transmission reached Earth on November 11, 1982. Controllers at NASA's Jet Propulsion Laboratory tried unsuccessfully for another six and a half months to regain contact with the lander, but finally closed down the overall mission on 21 May 1983.

Phobos 1

USSR Mars Orbiter/Lander—5000 kg—(July 7, 1988)
Phobos 1 was sent to investigate the Martian moon Phobos. It was lost en route to Mars through a command error on September 2, 1988.

Phobos 2

USSR Phobos Flyby/Lander—5000 kg—(July 12, 1988)
Phobos 2 arrived at Mars and was inserted into orbit on January 30, 1989. The orbiter moved within 497 miles (800km) of Phobos and then failed. The lander never made it to Phobos.

Mars Observer

USA Mars Orbiter—2573 kg—(September 25, 1992)
Communication was lost with Mars Observer on August 21, 1993, just before it was to be inserted into orbit.

Mars Global Surveyor

USA Mars Orbiter—1062.1 kg—(November 7, 1996)
Initiated due to the loss of the Mars Observer spacecraft, the Mars Global Surveyor (MGS) mission launched on November 7, 1996. MGS has been in a Martian orbit, successfully mapping the surface since March 1998.

Mars 96

Russia Orbiter & Lander—6200 kg—(November 16, 1996)
Mars '96 consisted of an orbiter, two landers, and two soil penetrators that were to reach the planet in September 1997. The rocket carrying Mars 96 lifted off successfully, but as it entered orbit the rocket's fourth stage ignited prematurely and sent the probe into a wild tumble. It crashed into the ocean somewhere between the Chilean coast and Easter Island. The spacecraft sank, carrying with it 9½ ounces (270g) of plutonium-238.

Mars Pathfinder

USA Lander & Surface Rover—870 kg—(December 1996)
The Mars Pathfinder delivered a stationary lander and a surface rover to the Red Planet on July 4, 1997. The six-wheel rover, named Sojourner, explored the area near the lander. The mission's primary objective was to demonstrate the feasibility of low-cost landings on the Martian surface. This was the second mission in NASA's low-cost Discovery series. After great scientific success and public interest, the mission formally ended on November 4, 1997, when NASA ended daily communications with the Pathfinder lander and Sojourner rover.

Nozomi

Japan Mars Orbiter—536 kg—(July 3, 1998) (Planet B)
Japan's Institute of Space and Astronautical Science (ISAS) launched this probe on July 4, 1998 to study the Martian environment. This would have been the first Japanese spacecraft to reach another planet. The probe was due to arrive at Mars in December of 2003. After revising the flight plan due to earlier problems with the probe, the mission was abandoned on December 9, 2003 when ISAS was unable to communicate with the probe in order to prepare it for orbital insertion.

Mars Climate Orbiter

USA Orbiter—629 kg—(December 11, 1998) (Mars Surveyor '98 Orbiter)
This orbiter was the companion spacecraft to the Mars Surveyor '98 Lander, but the mission failed.

Mars Polar Lander

USA Lander—583 kg—(January 3, 1999) (Mars Surveyor '98 Lander)
The Polar Lander was scheduled to land on Mars on December 3, 1999. Mounted on the cruise stage of the Mars Polar Lander were two Deep Space 2 impact probes, named Amundsen and Scott. The probes had a mass of 3.572 kg each. The cruise stage was to separate from the Mars Polar Lander, and subsequently the two probes were to detach from the cruise stage. The two probes planned to impact the surface 15–20 seconds before the Mars Polar Lander was to touch down. Ground crews were unable to contact the spacecraft, and the two probes. NASA concluded that spurious signals during the lander leg deployment caused the spacecraft to think it had landed, resulting in premature shutdown of the spacecraft's engines and destruction of the lander on impact.

2001 Mars Odyssey

USA Mars Orbiter and Lander/Rover—376.3 kg—(April 7, 2001) (Mars Surveyor 2001 Orbiter)
This Mars orbiter reached the planet on October 24, 2001 and served as a communications relay for future Mars missions. In 2010 Odyssey broke the record for longest-serving spacecraft at the Red Planet. It will support the 2012 landing of the Mars Science Laboratory and surface operations of that mission.

Mars Express

European Space Agency Mars Orbiter and Lander—666 kg—(June 2, 2003)
The Mars Express Orbiter and the Beagle 2 lander were launched together on June 2, 2003. The Beagle 2 was released from the Mars Express Orbiter on December 19, 2003. The Mars Express arrived successfully on December 25, 2003. The Beagle 2 was also scheduled to land on December 25, 2003; however, ground controllers have been unable to communicate with the probe.

Spirit (MER-A)

USA Mars Rover—185 kg—(June 10, 2003)
As part of the Mars Exploration Rover (MER) Mission, "Spirit," also known as MER-A, was launched on June 10, 2003 and successfully arrived on Mars on January 3, 2004. The last communication with Spirit occurred on March 22, 2010. JPL ended attempts to re-establish contact on May 25, 2011. The rover likely lost power due to excessively cold internal temperatures.

Opportunity (MER-B)

USA Mars Rover – 185 kg—(July 7, 2003)
"Opportunity," also known as MER-B, was launched on July 7, 2003 and successfully arrived on Mars on January 24, 2004.

Mars Reconnaissance Orbiter

USA Mars Orbiter—1031 kg—(August 12, 2005)
The Mars Reconnaissance Orbiter (MRO) was launched on August 12, 2005 for a seven-month voyage to Mars. MRO reached Mars in March 10, 2006 and began its scientific mission in November 2006.

Phoenix

USA Mars Lander—350 kg—(August 4, 2007)
The Phoenix Mars Lander was launched on August 4, 2007 and landed on Mars on May 25, 2008. It is the first in NASA's Scout Program. Phoenix was designed to study the history of water and habitability potential in the Martian arctic's ice-rich soil. The solar-powered lander completed its three-month mission and kept working until sunlight waned two months later. The mission was officially ended in May 2010.

Phobos-Grunt

Russia Mars Lander—730 kg/Yinghuo-I—China Mars Orbital Probe—115 kg—(November 8, 2011)
The Phobos-Grunt spacecraft was meant to land on the Martian moon Phobos. The Russian spacecraft did not properly leave Earth's orbit to set out on its trajectory toward Mars. Yinghuo-I was a planned Chinese Mars orbital probe launched along with Phobos-Grunt. Both craft were destroyed on re-entry from Earth orbit in January 2012.

Mars Science Laboratory

USA Mars Rover—750 kg—(November 26, 2011)
The Mars Science Laboratory was launched on November 26, 2011. With its rover named Curiosity, NASA's Mars Science Laboratory mission is designed to assess whether Mars ever had an environment able to support small life forms called microbes. Curiosity landed successfully in Gale Crater at 1:31am EDT on August 6, 2012.

Mars Orbiter Mission (Mangalyaan)

India Mars Orbiter—15 kg—(November 5, 2013)
The Indian Mars Orbiter Mission was launched on November 5, 2013, from the Satish Dhawan Space Center. It was inserted into orbit around Mars on September 24, 2014 and completed its planned 160-day mission duration in March 2015. The spacecraft continues to operate, mapping the planet and measuring radiation.

MAVEN

USA Mars Orbiter—2550 kg—(Launch November 18, 2013)
MAVEN (Mars Atmospheric and Volatile EvolutioN) was the second mission selected for NASA's Mars Scout program. It launched on November 18, 2013 and entered orbit around Mars on September 21, 2014. MAVEN's mission is to obtain critical measurements of the Martian atmosphere to further understanding of the dramatic climate change that has occurred over the course of its history.

InSight

USA Mars Lander—(Launch Window March 8—March 27, 2016)
InSight (Interior Exploration using Seismic Investigations, Geodesy and Heat Transport) is the twelfth mission in NASA's series of Discovery-class missions. InSight will take the first look into the deep interior of Mars to see why the Red Planet evolved so differently from Earth as one of our solar system's rocky planets. .

Courtesy of *http://history.nasa.gov/marschro.htm*

GLOSSARY

Caltech—the California Institute of Technology—or "Caltech"—is a private research university located in Pasadena, California, United States.

Compact Reconnaissance Imaging Spectrometer for Mars (CRISM)—the Compact Reconnaissance Imaging Spectrometer for Mars is a visible-infrared spectrometer aboard the Mars Reconnaissance Orbiter searching for mineralogic indications of past and present water on Mars.

De revolutionibus orbium coelestium—the seminal work on the heliocentric theory of the Renaissance Polish astronomer Nicolaus Copernicus.

Gas chromatograph—an analytical instrument that measures the content of various components in a sample. The analysis performed by a gas chromatograph is called gas chromatography.

ICBM—an intercontinental ballistic missile (ICBM) is a guided ballistic missile with a minimum range of 3400 miles (5500km) primarily designed for nuclear weapons delivery (delivering one or more thermonuclear warheads).

Jet Propulsion Laboratory (JPL)—a federally funded research and development center and NASA field center located in La Cañada Flintridge, California and Pasadena, California, United States.

Magnetometer—a measurement instrument used for two general purposes: to measure the magnetization of a magnetic material such as a ferromagnet, or to measure the strength and, in some cases, the direction of the magnetic field at a point in space.

Mars Orbital Camera (MOC)—a precision device built for NASA by Malin Space Science Systems in San Diego, California.

Mars Orbital Laser Altimeter (MOLA)—an instrument aboard the Mars Global Surveyor, which used infrared laser beams to map exact distances from the orbiter to the Martian surface in precise detail.

Marsquake—a quake on the planet, perhaps caused by land tides or volcanic eruptions.

Mass spectrometer—an instrument that can measure the masses and relative concentrations of atoms and molecules. It makes use of the basic magnetic force on a moving charged particle.

Micrometeoroid detector—an instrument designed for detecting tiny particles of rock in space, usually weighing less than a gram.

NASA—the National Aeronautics and Space Administration (NASA) is the United States government agency responsible for the civilian space program, as well as aeronautics and aerospace research.

Plutonium-238—a radioactive isotope of plutonium that has a half-life of 87.7 years.

Rock Abrasion Tool—a powerful grinder, able to create a hole 2in (45mm) in diameter and 0.2in (5mm) deep into a rock on the Martian surface.

Seismometer—a seismograph equipped for measuring the direction, intensity, and duration of earthquakes by measuring the actual movement of the ground.

Shallow Sub-surface Radar (SHARAD)—a device that looks for liquid or frozen water in the first few hundreds of feet (up to 1km) of Mars' crust.

Solar plasma probe—deployed to study radio and plasma that occur in the solar wind and in the Earth's magnetosphere before the solar wind.

Spectroscopy—the branch of science concerned with the investigation and measurement of spectra produced when matter interacts with or emits electromagnetic radiation.

Thermal Emission Spectrometer (TES)—TES collected two types of data, hyperspectral thermal infrared data from 6 to 50 micrometers (μm) and bolometric visible-NIR (0.3 to 2.9 μm) measurements.

Ultrastable Oscillator for Doppler Measurements (USORS)—a very precise clock that linked in to the radio on the spacecraft, with which scientists could track minute changes in the radio signal as it orbited Mars, allowing for extremely accurate measurements of the planet's gravitational field.

INDEX

ACKNOWLEDGMENTS

The list of people contributing to this book is long, and I am indebted to all of them.

Special thanks to Alison Moss and Anna Marx, exemplary editors at Carlton Books, for shepherding the book through the creative process and making it far better than it would have been in less capable hands. Piers Murray Hill, Editorial Director, deserves credit for championing quality books on space exploration and science. James Pople created the breathtaking layouts.

Jim Green, the Planetary Science Division Director at NASA, was kind enough to provide a wonderful introduction for the book. His untiring efforts to continue NASA's push into the solar system deserve high praise.

My agent John Willig is simply tops in the business, deeply engaged in every project and helpful beyond the call of duty. Susan Holden Martin, to whom this book is dedicated, is a longtime friend and stalwart ally in my professional life, as she has been to so many causes seeking to advance the exploration of Mars. Many of the non-governmental organizations that seek to further our efforts to explore Mars owe her a deep debt of gratitude for her tireless behind-the-scenes contributions.

No book of this type could be written without the unstinting assistance of the Jet Propulsion Laboratory. Their Office of Communication and Education, as capably directed by Blaine Baggett, provides resources that are simply unavailable elsewhere. Julie Cooper of the JPL Archives was incredibly helpful and provided a rare glimpse into JPL's history. Guy Webster, who oversees media outreach for Mars, was amazing to work with and is incredibly devoted to deepening the public understanding of NASA's exploration of Mars.

Rachel Tillman, the Director of the Viking Mars Missions Education and Preservation Project, and was unstinting in her efforts to provide unique and previously unpublished material for the book, and was a generous contributor to the accuracy of the Viking chapters. She has also been the longtime keeper of the flame for the Viking legacy, and through her efforts has preserved and protected a rare Viking lander assembly for public display. Her online museum, devoted to the Viking missions, can be seen at http://www.thevikingpreservationproject.org.

Rob Manning is JPL's Mars Engineering Manager and was the Chief Engineer for MSL and Mars Pathfinder, and led the system and EDL teams for MER. Rob contributed valuable documents, advice and an inside voice to the book. I swear, he must have a nuclear power supply like the Curiosity rover—he simply never runs short of energy.

Other folk from JPL and their affiliates who generously contributed to the "Insider's Voice" content include:

Ashwin Vasavada, John Grotzinger, John Casani (a legendary figure in planetary exploration), Peter Smith, Robert Leighton, Jeffrey Plaut, Steven Squyres, Richard Zurek, and Tom Rivellini. Their titles are noted in the special content they contributed in the appropriate chapters.

Buzz Aldrin provided access to his materials regarding the Mars Cycler, a visionary design for a permanent Mars infrastructure. Robert Zubrin, co-founder of the Mars Society and another visionary, supplied additional advice and materials.

Sam Teller and Phil Larson of SpaceX were patient with my questions; my thanks to you both. The work your company is doing towards reaching Mars enriches us all. Leonard David, probably the premiere space journalist (and now author) of our time, answered questions that few others could have. My deepest thanks.

Organizations that provided invaluable visual content include:

NASA/JPL, Caltech, the European Space Agency, The Indian Space Research Organization, Roscosmos, and the people who generously and selflessly provide materials to the various Wikipedia entries regarding Mars exploration.

A special thanks to Ken Kremer, a man of many talents including fine space journalism, who devotes countless hours to improving visuals that come from NASA's planetary missions. Some of the most amazing images you will see from JPL's Mars missions have been lovingly improved by his touch, and we are all richer for his efforts. His work can be seen at www.kenkremer.com and Universe Today online.

Also deserving of special thanks is David Clow, also a writer of exemplary space journalism and history, who selflessly went through the entire manuscript and provided carefully researched notes and insightful comments. He's a talented and gifted writer, and a tireless promoter of space exploration.

Of course, any errors or omissions that remain in the book are of my own doing. Research into space exploration history can be tricky, and even primary references frequently disagree. One makes the most informed choices possible, based on the available resources, but errors do occasionally creep in and they are an occupational hazard of writing in this genre.

Many thanks to all who participated, and we will share a toast when the first human footsteps are taken on Mars.

CREDITS

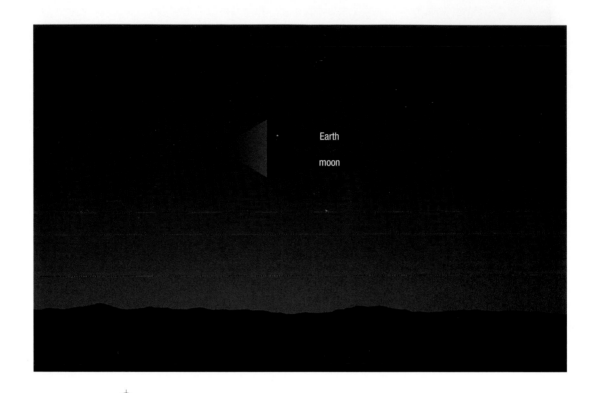

ABOVE: A view of the sky and Martian horizon taken by NASA's Curiosity rover in January 2014. The brightest point in the sky is Earth, and our moon is just below it.

The publishers would like to thank the following sources for their kind permission to reproduce the pictures in this book.

Key: t – top, b – bottom, r– right, l – left, c – center

Alamy Stock Photo/Heritage Image Partnership Ltd: 11

Bridgeman Images: 10tr, 10cr (Egypt), 10br

CIT/OSU/NASA: 180

Don Davis for NASA: 39t

ESA: 83l, 84, 85t, 85b, 86, 87cl, 87cr, 87b, 164, 165, 166tl, 166tr, 169; /Medialab: 82, 87t, 183

ESO/G. Hüdepohl: 168

ISRO: 149t, 149b, 150t, 151tl, 151tr

NASA: 40r, 172b, 173t, 173b, 174tl, 174tr, 175, 176–7, 179t, 179b, 181; /ESA/Medialab: 83r; /JPL-Caltech: 2, 7, 9, 16, 17, 18, 19, 20, 21, 22, 23, 25 (reprocessed by Ted Stryk), 26, 27, 29, 30, 32t, 32b, 33, 34l, 35t, 35b, 36t, 37, 39b, 40l, 41t, 41b, 43t, 43b, 44r, 45l, 45r, 48, 50t, 50c, 50b, 52, 53, 54t, 54b, 55t, 55b, 56, 57t, 57b, 58, 59t, 59b, 60l, 60r, 61, 62–3, 64tl, 64bl, 64–5, 65br, 66, 70, 75, 76t, 76b, 77b, 78, 79, 88, 89b, 90 tl and inset, 91, 92, 93t, 93b, 94t, 96–7, 100t, 100b, 106tl, 108, 109, 114b, 117bl, 117br, 118t, 119, 120b, 125br, 126, 127tr, 128tl, 128tr, 129tl, 129tr, 129bl, 137t, 137b, 139t, 141bl, 144tl, 144tr, 150b, 151b, 152, 153t, 153b, 154tl, 154bl, 154–5, 157, 158, 159l, 159r, 160, 162t, 162b, 163, 172t; /JPL-Caltech/ARC: 127tl; /JPL-Caltech/Arizona State University: 38, 80t, 80b, 97l; /JPL-Caltech/Cornell: 94b, 95, 97r, 98t, 98b, 99t, 101t, 101b, 104–5, 107t, 107b; /JPL-Caltech/GSFC: 41c, 46, 47, 73b, 89r, 117t, 125t; /JPL-Caltech/INL: 125bl; /JPL-Caltech/Lockheed-Martin: 74, 77t; /JPL-Caltech/MSSS: 67, 68–9, 73tl, 73tr, 73cl, 73cr, 81t, 81b, 131, 132–3, 134–5, 135b, 136t, 136b, 140–1t, 141br, 142–3t, 142b, 144t, 144b, 146, 147; /JPL-Caltech/MSSS/TAMU: 192; /JPL-Caltech/University of Arizona: 102, 103, 105t, 106tr, 111t, 111b, 112, 113t, 113cl, 113br, 114t, 115, 118b, 120t, 121, 122tl, 122tr, 122b, 144b; /JPL-Caltech/SwRI: 139b; /JPL/USGS: 31b, 166b, 170–1; /Lewis: 174bl; /NSSDC/GSFC: 36b; /Pat Rawlings, SAIC: 184–5

Powerhouse Museum: 31t

Private Collection: 10bl, 12, 14, 15t

Science Photo Library/Deltev Van Ravenswaay: 15b

Shutterstock.com: 24

SpaceX: 182

USPS: 34r

The Viking Mars Missions Education and Preservation Project (VMMEPP): 42, 44l, 49, 51

Wikimedia Commons: 10cr (Greece), 28, 138, 167

Every effort has been made to acknowledge correctly and contact the source and/or copyright holder of each picture and Carlton Books Limited apologizes for any unintentional errors or omissions, which will be corrected in future editions of this book.